METHODS OF MICROARRAY DATA ANALYSIS

Papers from CAMDA '00

METHODS OF MICROARRAY DATA ANALYSIS

Papers from CAMDA '00

edited by

Simon M. Lin

and

Kimberly F. Johnson

Duke University Medical Center

SPRINGER SCIENCE+BUSINESS MEDIA, LLC

Library of Congress Cataloging-in-Publication Data

A C.I.P. Catalogue record for this book is available
from the Library of Congress.

Originally published by Kluwer Academic Publishers in 2002
Softcover reprint of the hardcover 1st edition 2002

Printed on acid-free paper.

ISBN 978-1-4613-5281-5 ISBN 978-1-4615-0873-1 (eBook)
DOI 10.1007/978-1-4615-0873-1

Contents

Best Presentation – CAMDA '00

Quality Analysis and Data Normalization of Spotted Arrays

Feature Selection, Dimension Reduction, and Discriminative Analysis

Contributors

Aris, Virginie, Center for Applied Genomics, Public Health Research Institute, and Center for Computational Biology and Bioengineering, NJIT.

Berrar, Daniel, German Cancer Research Center, Intelligent Bioinformatics Systems Group, Heidelberg, Germany.

Cherry, Michael, Department of Genetics, Stanford University, Stanford CA.

Cho, Dong-Yeon, Biointelligence Laboratory, School of Computer Science and Engineering, Seoul National University, Seoul, Korea.

Darden, Thomas, Laboratory of Structural Biology, National Institute of Environmental Health Sciences, Research Triangle Park, NC.

Dubitzky, Werner, German Cancer Research Center, Intelligent Bioinformatics Systems Group, Heidelberg, Germany.

Ewing, Rob, Department of Plant Biology, Carnegie Institution of Washington, Stanford CA.

Finkelstein, David, Department of Plant Biology, Carnegie Institution of Washington, Stanford CA.

Gollub, Jeremy, Department of Plant Biology, Carnegie Institution of Washington, Stanford CA.

Grant, Gregory, Center for Bioinformatics, University of Pennsylvania (PCBI).

Granzow, Martin German Cancer Research Center, Intelligent Bioinformatics Systems Group, Heidelberg, Germany.

Hardy, Sarah, Program in Bioinformatics, NC State University, Raleigh, NC.

Hwang, Kyu-Baek, Biointelligence Laboratory, School of Computer Science and Engineering, Seoul National University, Seoul, Korea.

Kim, Sung-Dong, Biointelligence Laboratory, School of Computer Science and Engineering, Seoul National University, Seoul, Korea.

Li, Leping, Biostatistics Branch and Laboratory of Structural Biology, National Institute of Environmental Health Sciences, Research Triangle Park, North Carolina .

Li, Wentian, Laboratory of Statistical Genetics, The Rockefeller University, New York, NY.

Lu, Jun, Program in Bioinformatics, NC State University, Raleigh, NC.

Manduchi, Elisabetta, Center for Bioinformatics, University of Pennsylvania (PCBI).

Moore, Jason, Program in Human Genetics, Department of Molecular Physiology and Biophysics, Vanderbilt University Medical School, Nashville, TN.

Muse Spencer, Program in Bioinformatics, NC State University, Raleigh, NC.
Nguyen, Danh, Center for Image Processing and Integrated Computing, University of California, Davis, CA.
Page, Grier, Department of Biometry and Epidemiology, Medical University of South Carolina, Charleston, SC.
Park, Sang-Wook, Biointelligence Laboratory, School of Computer Science and Engineering, Seoul National University, Seoul, Korea.
Parker, Joel, Program in Human Genetics, Department of Molecular Physiology and Biophysics, Vanderbilt University Medical School, Nashville, TN.
Pedersen, Lee, Department of Chemistry, University of North Carolina, Chapel Hill, NC.
Recce, Michael, Center for Applied Genomics, Public Health Research Institute, and Center for Computational Biology and Bioengineering, NJIT.
Rocke, David, Center for Image Processing and Integrated Computing and Department of Applied Science, University of California, Davis, CA.
Somerville, Shauna, Department of Plant Biology, Carnegie Institution of Washington, Stanford CA.
Spruill, Susan, PPGx. Inc., a subsidiary of DNA Sciences, RTP, NC.
Sterky, Fredrik, Department of Plant Biology, Carnegie Institution of Washington, Stanford CA.
Stoeckert, Jr., Christian, Center for Bioinformatics, University of Pennsylvania (PCBI).
Tao, Wen-Li, Program in Bioinformatics, NC State University, Raleigh, NC.
Weinberg, Clarice, Biostatistics Branch and Laboratory of Structural Biology, National Institute of Environmental Health Sciences, Research Triangle Park, NC.
Weir, Bruce, Program in Bioinformatics, NC State University, Raleigh, NC.
Yang, Yaning, Laboratory of Statistical Genetics, The Rockefeller University, New York, NY.
Zhang, Byoung-Tak, Biointelligence Laboratory, School of Computer Science and Engineering, Seoul National University, Seoul, Korea.
Zhang, Hong, BIOwulf Technologies, Savannah, GA.
Zhang, Zhen, Department of Biometry and Epidemiology, Medical University of South Carolina, Charleston, SC and 3Z Informatics, Mt. Pleasant, SC.

Acknowledgements

The editors would like to thank the contributing authors for their fine work. We would also like to thank Anna Menzies and Patrick McConnell for their assistance in preparing this volume. Our thanks to the advisory committee, the scientific committee and the additional reviewers needed to make CAMDA'00 an outstanding conference. We would also like to thank our supporters at Duke University; Michael Colvin, Director of the Duke Comprehensive Cancer Center, the Center for Bioinformatics and Computational Biology, John Harer, Vice Provost for Academic Affairs, and the Duke Specialized Project for Research Excellence (SPORE) in Breast Cancer. Of course the conference would not be possible without the generous support of our corporate sponsors including The Scientist, Lab-on-a-Chip.com, The Research Place, and Gene Tec Corporation. In addition, we gratefully acknowledge The North Carolina Biotechnology Center for providing a generous meeting grant.

Preface

The advent of microarray technology has produced a phenomenal growth in the amount of data available to researchers seeking to understand the complexities of gene function. This multitude of data offers the opportunity for expanded insight into genomics, but presents new research problems. Now that we have all this data, how do we analyse it? This question is at the heart of bioinformatics and indeed at the heart of this volume as the research community begins its search for answers.

In December of 2000, a conference on the Critical Assessment of Microarray Data Analysis (CAMDA) was held to stimulate discussion of the state-of-the-art in this emerging field. Of the thirty-three papers submitted, fourteen were presented orally and all were presented as posters. This volume highlights ten oral presentations, preceded by two papers that provide an overview of selected microarray analysis methods.

As editors, we have not comprehensively edited these papers, but have provided comments to encourage clarity and expansion of ideas. Each paper was peer-reviewed and submitted back to the author for further revisions. The editors do not propose these methods as the de facto standard for microarray analysis. There are no accepted protocols and it will be some time before a standard is established. The main focus of CAMDA is to highlight the many methods now being introduced, and to provide the opportunity for the scientific community to discuss and critique these methods.

As you read this volume, you may disagree with the methodologies presented or feel that you have a better model. If so, we invite you to submit your work and join us at the next CAMDA conference. Your work may bring us a step closer to discovering the secrets now hidden within the genome.

Simon Lin

Kimberly Johnson

INTRODUCTION

DNA microarrays are an exciting new development in the field of genomics research. For the first time, scientists have gained a more comprehensive view of the genome than had been possible in the past. Rather than investigating one or two genes, researchers can now view the expression of thousands of genes at one time. However, this giant step forward has also arrived with a proportionate set of problems, most notably the need for statisticians, biologists and computer scientists to unite to produce new methodologies to analyze the data.

A CRITICAL ASSESSMENT OF MICROARRAY DATA ANALYSIS TECHNIQUES

In early 2000, it was difficult to find information on analytical model for microarrays. Many conferences were devoted to general microarray issues such as array fabrication and quality control, but none were dedicated to analytical problems. Thus we decided to hold a Critical Assessment of Microarray Data Analysis (CAMDA) conference, modelled after the CASP experiment in protein modelling. In this approach, the entire scientific community would be invited to analyze the same data set and report their findings in an annual conference. Our goals were to answer two specific questions arising from microarray technology:

- Can traditional statistical approaches adequately handle the volume and complexity of the data now being produced?
- If new analytical methods are needed, how can they be appraised?

When we initially developed CAMDA, we thought it would be a small conference with perhaps fifty researchers. We were surprised by the strong response, eventually resulting in 250 attendees from seven countries. CAMDA'00 attracted a large number of submissions from researchers with very different backgrounds. The scientific committee selected fourteen contributions for oral presentation at the conference and all submissions were presented as posters. The thirty-three oral and poster presentations were complemented by opening remarks from John Weinstein (NIH), keynote lectures from Mike West (Duke University) and Gavin Sherlock (Stanford University), and closing remarks from Athel Cornish-Bowden (CNRS, France). At the end of the conference, attendees voted on the "Best Presentation" with the Scientific Committee providing weighted votes. After the meeting, the overwhelming response of attendees was that CAMDA provided an important forum for a meeting of the minds in this arena.

CAMDA 2000 DATA SETS

In order to create opportunities for a direct comparison of techniques, each contributor used one of two previously published data sets. The first data set was from the seminal work of Spellman et al. [*Mol Biol Cell* 1998, 9:3273-3279] for cataloguing cell-cycle regulated genes in yeast by cDNA microarrays. The second data set from Golub et al. [*Science* 1999, 286:531-537] utilized Affymetrix oligoarrays to profile leukemia patients. We chose these two data sets because they represented the two different types of microarrays, and because they represented two distinctly different application areas; using yeast to study a basic biological process and using leukemia as a model to study human cancer.

PAPERS PRESENTED IN THIS VOLUME

As a prelude to the methodology papers presented at CAMDA, we begin with two tutorials. One on data mining and machine learning techniques, and the other on special issues of evolutionary computation. These papers provide a general understanding of these techniques, some of which are presented in detail in the CAMDA papers.

The first CAMDA paper in this volume is titled "Using non-parametric methods in the context of multiple testing to determine differentially expressed genes" by Grant et al. The authors of this paper were presented with the CAMDA'00 Best Presentation Award. The paper addresses the issue of identifying the differentially expressed genes by a non-parametric approach called PaGE. PaGE also determines the confidence associated with these predictions.

The rest of the papers are organised by topic, although you will note a progression of methods that makes it difficult to draw the line between them at times. We start with a paper by Finkelstein et al. addressing normalization issues with spotted arrays. The next section presents six papers that highlight feature selection, dimension reduction, and discriminative analysis issues. Later in this section, however, you will begin to see the integration of elements of machine learning techniques in the analytical processes. Finally we present two papers dedicated to machine learning techniques as they are brought to bear on the conference data sets.

It has been clear for some time that data normalization is a fundamental issue in microarray analysis since almost all downstream algorithms depend on appropriate data preprocessing. What becomes clear in the presentation of these papers is that the feature selection and dimension reduction problem must be addressed before effectively building discriminative classifiers. Further, it was clear that most researchers feel new methods are needed to address the complexities of microarray data. However, several researchers presented methods using traditional statistical approaches that work equally well. Whether these approaches are adequate only for the leukemia data set remains to be seen. Many of the models could be effectively handled by high performance PCs, although some methods did require parallel computing.

Overall, an important point made repeatedly at the conference was the need to put these studies into the correct biological context. Even with many sophisticated methodologies available, if we don't apply these correctly, then we are doomed to failure. As Dr. Werner Dubitzky noted in his CAMDA presentation, "If the only tool you have is a hammer, then everything looks like a nail". The CAMDA conference showed us that we have more than hammers available.

WEB COMPANION

Additional information for many of these chapters can be found at the CAMDA website (http://camda.duke.edu), where links to many of the algorithms and other methods can be found. You can also find many of the conference slides, as well as information about future CAMDA conferences. As the CAMDA conference grows, we hope this website will become a community-wide resource for those looking for the latest in microarray techniques.

1

DATA MINING AND MACHINE LEARNING METHODS FOR MICROARRAY ANALYSIS

Werner Dubitzky, Martin Granzow, Daniel Berrar
German Cancer Research Center, Intelligent Bioinformatics Systems Group, Heidelberg, Germany

Abstract: Microarray experiments provide the scientific community with huge amounts of data. Without appropriate methodologies and tools, significant information and knowledge hidden in these data may not be discovered. Therefore, there is a need for methods capable of handling and exploring large data sets. The field of data mining and machine learning provides a wealth of methodologies and tools for analyzing large data sets. We review two classical machine learning techniques suitable for microarray analysis, namely *decision trees* and *artificial neural networks*. We outline how these approaches fit into a wider data mining framework.

Key words: Microarray analysis, decision tree, artificial neural network, data mining, machine learning

INTRODUCTION

The wealth of microarray data now available poses enormous challenges such as the analysis of images generated by microarray experiments [Chen et al., 1998], the study of the variability of gene expression patterns [Newton et al., 2001], and the reverse engineering of biochemical and signaling networks from microarray data [Voit & Radivoyevitch, 2000]. At present, most microarray analyses are concerned with (i) *cluster analysis*: the identification of new subgroups or classes of some biological entity (e.g., tumors), or (ii) *discriminant analysis*: the classification of entities into known classes (e.g., diseases, therapies). A typical characteristic of microarray data is the large number of variables (e.g., genes) relative to the number of observations (e.g., samples). For example, a study from 1999 analyzed gene expression data from 72 leukemia patients for class discovery (cluster analysis) and prediction (discriminant analysis) based on 7,070 genes [Golub et al., 1999]. Both the number of variables and the number of observations are expected to grow in the future.

Within the context of microarray data analysis, we outline data mining as a framework comprising methodologies and tools for analyzing large data sets including microarray data. We focus our discussion on two popular machine learning techniques for molecular classification of cancer and identification of potentially relevant genes. The techniques in question are *decision trees* and *artificial neural networks* [Dayhoff, 1996]. Decision trees belong to the class of so-called *symbolic* methods. The attractiveness of decision trees is largely due to their ability to express the learned models as symbolic rules that can readily be understood by humans. Neural network approaches, on the other hand, represent their learned knowledge as patterns of connectivity that exist among the nodes of the network. This type of knowledge representation is sometimes called *subsymbolic*, and is not readily intelligible to humans.

Many implementations of decision trees and neural networks exist. To make our discussion more concrete, we limit ourselves to two of the most popular algorithms currently available in proprietary machine learning and data mining software: the decision tree algorithm C5.0 [Quinlan & Quinlan, 1997] and the well-known backpropagation algorithm for neural networks [Witten & Frank, 2000; Berry & Linoff, 1997; Dayhoff, 1996]. In [Dubitzky et al., 2000] we have applied these two algorithms in a comparative study based on the previously mentioned work by Golub and colleagues [Golub et al., 1999]. The results our study are also presented in some detail within this volume.

Our key motivation is to contribute to the understanding of the underlying techniques, their benefits and their limitations within the context of microarray analysis. Besides delving into detail with regard to neural nets and decision trees, we also briefly discuss cross-validation and multiple-model strategies such as boosting. This entire discussion should not be seen as an exhaustive or complete treatment of these techniques.

DATA MINING AND GENERAL MOTIVATION

We are generally motivated by developing, validating, and applying *data warehousing* and *data mining* methodology and tools to life science problems. Therefore, we view the presented work as part of a wider framework. With the data analysis problem at hand in mind, we briefly outline related aspects of this framework.

To adequately organize and interpret the deluge of information generated by high-throughput technologies in molecular biology, the adaptation of existing and the development of new computational methodologies and tools are required. The principal approach to analyzing and interpreting biological data is to abstract them into logical structures that support and incrementally develop a more general conceptual framework for characterizing, explaining, and predicting processes in living systems. Among other critical technologies (e.g., agents, networking, knowledge-based systems) data mining is certainly one important element in the sought-after new computational framework.

Data mining is concerned with the automatic or semiautomatic exploration and analysis of large quantities of data in order to discover meaningful, previously unknown, non-trivial, and potentially useful knowledge (patterns, relationships, trends, rules, anomalies, dependencies) [Witten & Frank, 2000; Berry & Linoff, 1997]. Data mining is a continuous and iterative process. It involves the use of software, sound methodology, and human creativity to achieve new insight through the exploration of data. The goal of data mining is to allow the end-user (e.g., scientist, engineer, physician) to improve his or her decision-making. Compared with classical statistical approaches, data mining is perhaps best seen as a process that encompasses a wider range of integrated methodologies and tools including databases, network technology, modeling, algorithms, statistics, machine learning, knowledge-based techniques, uncertainty handling, and visualization.

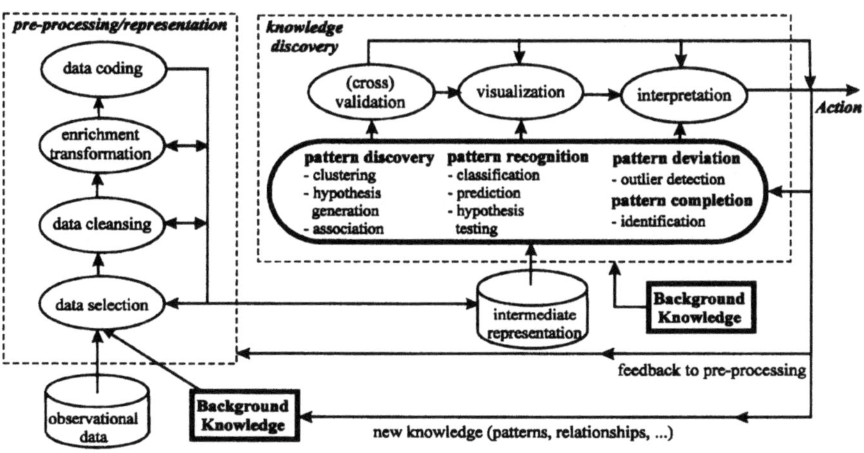

Figure 1. The data mining process.

The data mining process and some of its main elements and processes are depicted in Figure 1. In the diagram, processes are portrayed as ovals and rounded boxes.

The left side of the diagram illustrates the data access and data pre-processing or representation part of the overall process. This phase transforms the original data (from one or more sources) into an intermediate format. This process may be straightforward or arbitrarily complex, especially when complex background knowledge needs to be incorporated. Currently, it is being discussed if this process can be systematized by means of *data warehousing* methodology and technology [Dubitzky et al., 2001]. A major obstacle towards this goal is the complexity (logical/structural and temporal) and the global proliferation of the formal background knowledge in biology and life science.

The right part in Figure 1 shows the so-called *knowledge discovery* process. In this process, the pre-processed data is analyzed using a suitable pattern analysis methods. Which method is most suited for the problem at hand depends on the underlying domain, the question being asked, the structure of the available data, and the data mining *task* that is to be performed. Typical data mining tasks include *discriminant analysis*, *estimation/regression*, *clustering*, *association analysis* [Hipp et al., 2000], and *deviation detection* [Arning et al., 1996]. Two of the most common data mining tasks, namely discriminant analysis and clustering, are briefly discussed below. Most of the discussion below is presented from the perspective of discriminant analysis.

Both discriminant analysis and cluster analysis are often referred to as classification. However, cluster analysis is quite different from discriminant analysis in that it actually establishes groups (i.e., classes) of objects or entities, whereas discriminant analysis assigns objects to groups that were defined in advance. Clustering seeks a convenient and valid organization (and description) of data and not the generation of rules for separating future data into categories. Thus, cluster analysis is often used when a set of cases or observations is to be divided into "natural" groups. More specifically, clustering is a process or task that is concerned with establishing classes or groups from a set of observations, and with the definition or description of the classes that are identified. Because of this added requirement and complexity, clustering is considered a higher-level process than discriminant analysis. General definitions of both discriminant analysis (also called classification) and clustering (sometimes referred to as automatic classification or partitioning) are provided in Definition 1 and Definition 2 respectively.

Definition 1. *Discriminant analysis*: Given a universe, U, the subsets (predefined classes) $C_1, C_2, ..., C_n \subset U$, and the indices $1, 2, ..., i, ..., n \in I$, then a *classifier* is a function, $f(x)$: $U \to I$, such that $f(x) = i$ implies $x \in C_i$ (that is that x is a member of class C_i).

Discriminant analysis is often used to verify or disprove preconceived notions and ideas concerning the relationships in the data (*hypothesis testing*), or to *explain* or categorize a certain attribute or property, e.g., tumor type, of the analyzed observations.

Definition 2. *Cluster analysis*: Given a set X with n objects $X = \{x_1, x_2, ..., x_n\}$ with each object, x_i, $i = 1 ... n$, described by m attributes, $x_i = (x_{i1}, x_{i2}, ..., x_{im})$, determine a classification (grouping, clustering) that is most likely to have generated the observed objects. [Upal & Neufeld, 1996]

A frequently applied approach to clustering attempts to produce classes or groups that maximize similarity within classes but minimize similarity between classes. In the context of microarray data analysis, clustering methods may be useful for automatically detecting new subgroups (*hypothesis generation*) of clinical or biological entities (e.g., tumors, genetic risk groups) in the data [Granzow et al., 2001].

Data mining and data warehousing were first applied on a large scale in so-called customer-oriented businesses including the financial, marketing, retailing, and transportation sector. Typically, in these applications the number of variables is less than 100 and the number of observations ranges from several thousands to several millions. This situation is quite different for microarray data where the number of observations is often very small relative to the number of variables by a factor 10 to 100 or more. Because of this discrepancy it is mandatory to perform cross-validation procedures in order to obtain statistically reliable estimates for the performance of the derived models.

CROSS-VALIDATION

In the presence of a *large* number of observations, cross-validation procedures take up to 90% of the data to construct a discriminant model and estimate the model's future performance by validating it against the remaining observations. The data set for construction of the model is usually called the *training set*, and that for validation is referred to as the *validation set*. Given a large number of cases, this procedure yields a statistically sound estimate for gauging the expected performance of the model. However, when there is only a limited number of observations, this procedure is not very effective, as (1) the two data sets may be overly biased, and (2) not all available data is used for training.

For example, given a two class problem with a total of 100 observations and the absolute class frequencies of $F(\text{A}) = 80$, for class A, and $F(\text{B}) = 20$, for class B, a randomly sampled training set of size 75 may contain no instance of class B at all. Thus, the construction of a good model from this strongly "biased" training set will be difficult. Also, in this example, 25% of the available data is not used for model construction.

A way to avoid these problems is n-fold cross-validation. Generally, n-fold cross-validation divides the available data set into n *folds* or subsets. Based on these n data sets, n different models are developed, each using n-1 data sets for training and the remaining data set for validation. The advantage of cross-validation is that it increases the number of data patterns available for training, while using every data pattern available for testing. The diagram in Figure 2 depicts the general approach to cross-validation. In the diagram, the index $c = 1, 2, ..., i, ..., n-1, n$, refers to the cross-validation fold, and $f_i(x)$ denotes the model resulting from the ith cross-validation iteration or fold. For each cross-validation fold, the sets S_1 to S_n form a mathematical *partition* of the original data set.

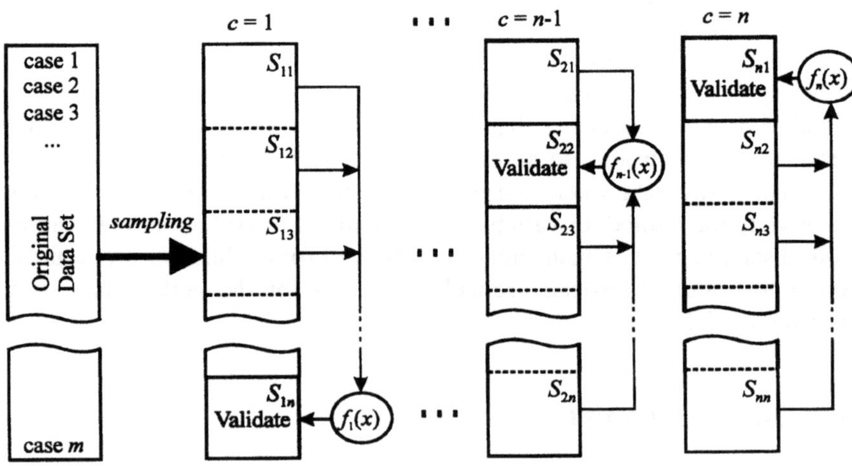

Figure 2. General model for cross-validation.

Once all n cross-validation iterations have been performed, one can estimate the future performance of the general approach by averaging the performances of the n models on the corresponding validation sets. From there one may proceed by either selecting the best-performing individual model or by constructing the final model using all the available data.

It is important to understand that in the standard cross-validation procedure repeated *uniform* sampling *without replacement* is performed. Here, uniform means that each "current" case in the original data set has the same probability of being drawn, and without replacement refers to the fact that once a case is assigned a set (training or validation set) it cannot be assigned or drawn into another set within the same cross-validation fold.

MULTI-MODEL APPROACHES

The instability of a prediction method refers to the sensitivity of the final model to small changes in the training set with regard to the prediction accuracy of the model. Small changes in the learning set may lead to large changes in the prediction performance. Typical unstable machine learning methods are decision trees, while, for example, *k*-nearest neighbor and neural models belong to the class of stable methods. An approach to address the instability problem is to construct multiple models and combine them to make up the final model. There exist three commonly used methods for constructing multi-model systems: *boosting*, *bagging*, and *stacking* [Ng, 1996]. These could be viewed as representatives of a more general framework called *data and information fusion* [Azuaje et al., 1999]. Boosting, bagging, and stacking deal with creating and combining multiple classifiers but differ in how the classifiers are trained and in how their outputs are combined. Such models often improve accuracy by focusing the learning process on examples in the data that are harder to model than others. We have applied the C5.0 decision algorithm using its boosting option to analyze microarray data in [Dubitzky et al., 2000]. In the following we briefly describe the basic concepts of boosting.

Boosting is an algorithm that creates several different models and combines their predictions using a weighted voting scheme (e.g., plurality voting). The elements of a boosting setup and their relationships are depicted in Figure 3. In boosting procedure, *N* different (perturbed) training set replicas are sampled adaptively (with *non*-uniform sampling probabilities and *with* replacement) from the learning set (black box in Figure 3. The predictions of the combined model are generated by means of a weighted voting scheme, where each individual prediction model carries a different weight.

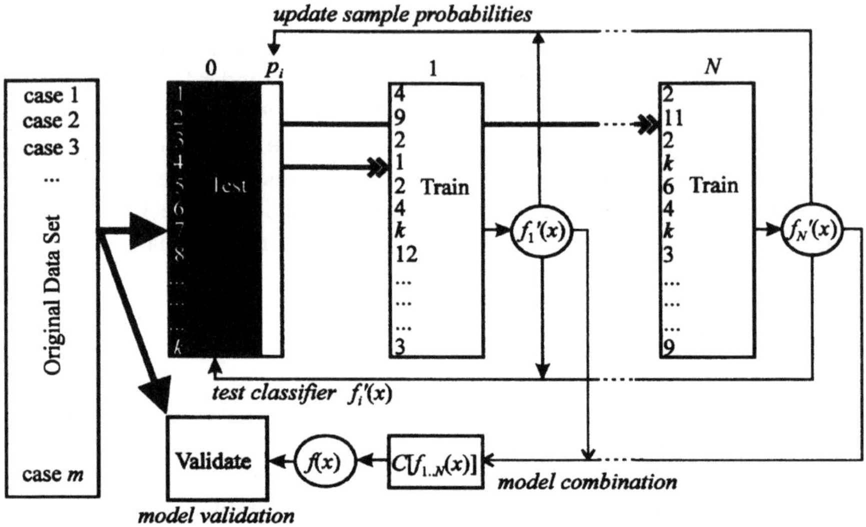

Figure 3. Boosting (thick arrow = uniform sampling, *without* replacement; double arrow = non-uniform sampling *with* replacement using sampling probabilities depicted by p_i).

The adaptive sampling procedures increase the probability of an instance to be sampled based on the performance of the classifier in the previous iteration. Instances that were most often misclassified are assigned an increased probability for being sampled in the next round. The general algorithm for boosting is governed by the following scheme (which is also illustrated in the diagram in Figure 3.

Step 1: Use current (index i) sampling probabilities, p_i, and sample with replacement k instances from the learning set (black box labeled Test in the diagram).

Step 2: Build current classifier, $f'_i(x)$, based on the current training set.

Step 3: Establish current classifier's performance by testing it against the learning set (black box) and keep a record of correctly/incorrectly classified cases, and

Step 4: Update the sampling probabilities (feedback arrow in upper part of diagram) for each instance based on the classifier's performance and calculate a weight for the classifier also based on its performance (illustrated by the small box in the lower part of the diagram).

Step 5: Stop if number of classifiers is reached, otherwise go to Step 1.

In a microarray analysis comprising a total of 72 samples (learning set 27 ALL, 11 AML; and test set of 20 ALL 14 AML) and the expression levels of 7,070 genes, we showed that the C5.0 decision tree algorithm with boosting (10-fold and 20-fold boosting) was superior to the non-boosting version [Dubitzky et al., 2000]. Over a 6-fold cross-validation procedure the non-boosting experiment achieved an average classification accuracy of 84.09%, with 10-fold boosting 91.87%, and with 20-fold boosting 92.98%.

Decision Trees

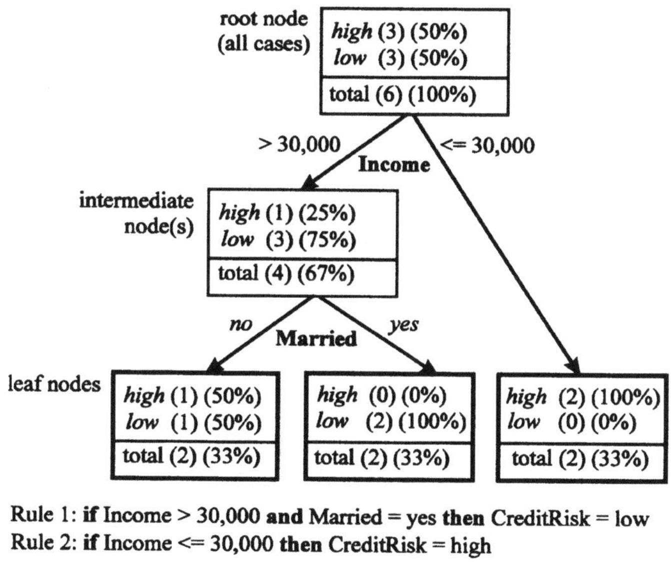

Figure 4. Example of simple decision tree after learning.

So-called *non-parametric* models rely heavily on the empirical analysis of large data sets rather than on prior domain knowledge, i.e., *domain-specific knowledge* about the input-output relationship of the underlying system or problem. The fundamental assumption of non-parametric models is that the consistently observed *relationships* or *patterns* in large data sets will recur in future observations. This assumption is crucial as the model attempts to form or learn generalizations based on the presented data. The advantage of non-parametric models is that (1) they do not require a thorough understanding of the underlying system or problem, and (2) they can be used to build arbitrarily complex models, that are highly non-linear and not restricted by human comprehension. Typical non-parametric approaches include decision trees, neural networks, genetic algorithms, and nearest neighbor methods.

A decision tree is an approach for generating models that are both descriptive and predictive. The term decision tree is derived from the presentation of the resulting model as a tree-structure (Figure 4 illustrates a very simple decision tree).

The example illustrated in Figure 4 shows data (six observations in total: 3 high credit risk, and 3 low credit risk) from a credit risk study involving factors like marital status (Married) and annual earnings (Income), and a decision variable (CreditRisk). The *root* node at the top of the tree represents all observations. Each *intermediate* and *leaf node* contains information about the number of cases at that node and the distribution of the independent variable. Leaf nodes give the classification that applies to all instances that reach the leaf (leaf nodes may not be "pure"). Nodes in a decision tree involve a decision or test about the attributes or variables that reach the node. Typically, this decision is a statistical test on how well an attribute alone classifies the training samples. During training the best attribute is selected and used at this node to evaluate future samples. Also, a descendant of the node is created for each possible value of the selected attribute (in case of discrete variables), unless the node is a leaf node. Continuous variables are first categorized into ranges before branching off a subnode. For example, at the root node in Figure 4 the attribute Income was found to be the best, and the statistical test determined that a split at 30,000 partitioned the underlying cases most effectively. Before we turn to a typical statistical test used to select attributes and generate splits (i.e., create subnodes), we briefly outline the general algorithm used for constructing decision trees.

Decision tree learning follows a kind of top-down, divide-and-conquer strategy. The basic algorithm for decision tree learning can be described as follows:

1. Select (based on some measure of "purity" such as entropy, information gain, or diversity) an attribute to place at the root of the tree and branch for each possible value of the tree. This splits up the underlying case set into subsets, one for every value of the considered attribute.
2. Recursively repeat this process for each branch, using only those cases that actually reach that branch. If at any time all instances at a node have the same classification, stop developing that part of the tree.

Typical statistical measures used in decision tree algorithms to select an attribute include *information gain* measures (as implemented in the C5.0 algorithm), and *diversity measures* (as implemented in the CART algorithm) [Mitchell, 1997; Breiman et al., 1984]. Here, we illustrate the information gain measure. To define information gain, we first begin by defining another measure, called *entropy*, that characterizes the (im)purity of a collection of observations.

Definition 3. Given a set of predefined classes, $C = \{c_1, c_2, ..., c_n\}$, and the indices $1, 2, ..., i, ..., n \in I$, and the (training) set X with m observations, $X = \{x_1, x_2, ..., x_m\}$, with each object, $x_j \in X$, described by k attributes and one class $c_j \in C$, such that $x_j = (x_{1j}, x_{2j}, ..., x_{kj}, c_j)$. Then the entropy, *entropy*(X), of the set X relative to this n-wise classification is defined as

$$entropy(X) = \sum_{i=0}^{n} - p_i \log_2 p_i \qquad [1]$$

where p_i is the proportion of X belonging to class $c_i \in C$.

The measure called *information gain*, *gain*(X,A), is simply the expected *reduction in entropy* caused by partitioning the set of observations, X, based on an attribute A (see equation [2]).

$$gain(X, A) = entropy(X) - \sum_{v \in values(A)} \frac{|X_v|}{|X|} entropy(X_v) \qquad [2]$$

where *values*(A) is the set of all possible (discrete) values of attribute A, and X_v is the subset of X for which attribute A has the attribute value v, i.e., $X_v = \{x \in X \mid A(x) = v\}$.

The strengths of decision trees are:

1. the ability to generate understandable knowledge structures, i.e., hierarchical trees or sets of rules;
2. a low computational cost when the model is being applied to predict or classify new cases;
3. the ability to handle symbolic and numeric input variables; and
4. provision of a clear indication of which attributes are most important for prediction or classification.

The weaknesses of decision trees include:

1. a limited ability to handle estimation or regression tasks where the goal is to predict the value of a continuous variable;
2. error-prone when the number of training examples per class is small;
3. a moderate to high computational cost for growing a decision tree;
4. decision trees generate rectangular classification boxes that may not correspond well with the actual distribution of the underlying regions in the decision space; and
5. without further transformation of the input data, decision trees are not very suitable for learning tasks involving sequential data such as time series.

Nowadays there is a plethora of decision tree algorithms available. Two of the most popular and widely used decision tree algorithms are the C5.0 (or its precursor C4.5) [Quinlan & Quinlan, 1997] and CART [Breiman et al., 1984]. Being very similar in many respects, the main differences of these two decision tree algorithms are:

1. The criteria they use to select the candidate attribute for splitting. As mentioned before, C5.0 uses a measure of information gain and CART uses a measure of diversity.
2. The way they split the tree at each node. This is perhaps the most important difference. C5.0 produces trees with varying numbers of branches per node, whereas the CART algorithm performs a binary split at each node and therefore always constructs a binary tree.
3. The pruning method for reducing the complexity of the constructed tree. Generally, CART makes reference to unseen data in order to prune the tree, whereas C5.0 does not consult data beyond the training set.

To decide which of the two decision trees to use depends on the underlying task and data. Clearly, an insistence on binary splits (as it is the case with CART), on a variable that is inherently more various, leads to unnecessary complexity in the generated tree. High branching factors (generated by C5.0), on the other hand, lead to a quick reduction of the population of training observations available at each node at lower levels in the tree. This makes further splits less reliable.

Neural Networks

Artificial neural networks are an emerging computational technology which can significantly enhance a number of applications ranging from robust pattern detection over signal filtering and compression to adaptive control and optimisation tasks. The key properties of neural networks can be summarized as follows:

- *adaptive learning*: An ability to learn how to do tasks based on the data given for training;
- *self-organization*: During learning a neural network can create its own organization or representation of the information it receives;
- *fault tolerance*: Noisy data or a partial destruction of the network leads to the corresponding degradation, however, some of the network's capabilities may be retrained even with major damage; and
- *real-time operation*: Neural network computations may be carried out in parallel, and special hardware is being developed which take advantage of this capability.

The major strengths of neural networks include:

1. ability to handle a range of problem tasks including classification (discrete outputs) and estimation or regression tasks (continuous outputs);
2. ability to handle many interacting variables (e.g., images) and non-linear (input-output) behavior of the underlying phenomena;
3. ability to handle symbolic and numeric input variables;
4. provision of an indication (through sensitivity analysis) of which attributes are most important for prediction or classification; and
5. low to moderate computational cost when the model is being applied to predict or classify new cases.

The weaknesses of neural networks are:

1. the learned models are described by patterns of connectivity strengths, therefore the rules learned by neural networks cannot be made explicit in symbolic form (the net *is* the rule); in applications where mere

prediction/classification performance is not the only criterion this may be a severe problem;
2. the need to preprocess the inputs in the range of [-1,1] or [0,1] (this is automatically done by most tools);
3. the risk of premature convergence to an inferior solution (this is normally addressed by performing a sensible cross-validation procedure);
4. depending on the chosen topology and the learning algorithm, there may be moderate to high computational costs involved in the learning process; and
5. difficulty of defining/finding the optimal network topology for a given task.

Artificial neural networks consist of basic units, called *processing elements* or *neurons*, that are modeled on biological neurons. Each processing element has many inputs that it combines into a single output. The learning capability of a neural network is defined by the processing function (called *activation function*) of the processing elements and the way these elements are connected together, i.e., the network's architecture or *topology*. Once a network has been trained, the learned knowledge is represented (stored) in the network by the *weights* connecting the processing elements.

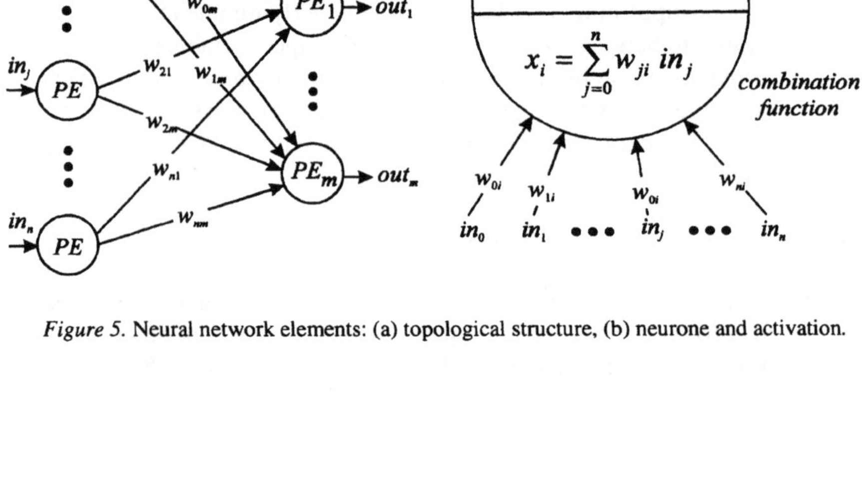

Figure 5. Neural network elements: (a) topological structure, (b) neurone and activation.

Figure 5 illustrates a simple neural network and its elements. Usually, a network of this type has m output processing elements, $PE_1, \ldots, PE_m$, each of which receives $n + 1$ inputs, namely, $in_0, in_1, in_2, \ldots, in_j, \ldots, in_n$ from the corresponding input processing elements; in_0 is a special input called *bias input*, it is always set to 1 (for the net in Figure 5-a: $m = 2, n = 3$). The bias input, in_0, together with the corresponding weights $w_{0i}, \ldots, w_{0m}$ represent thresholds used to control the behavior of the processing elements $PE_1, \ldots, PE_m$. The vector $\mathbf{C} = (in_1, in_2, \ldots, in_j, \ldots, in_n)$ represents an actual case that is to be processed. Each input of a processing element is associated with a set of weights. For example, processing element PE_1 is associated with the weights $w_{01}, w_{11}, \ldots, w_{j1} \ldots, w_{n1}$.

The activation function of each processing element or neurone consists of a *combination function* and a *transfer function* (see Figure 5-b). The combination function merges all inputs, and their weights, into a single value. The most common combination function is the *weighted sum* which is given by equation [3] below. The second part of the activation function is the transfer function. It transfers the output value of the combination function to the output of the processing element. Commonly used transfer functions are the *linear*, *hyperbolic tangent*, and *sigmoid* function; the *sigmoid* function is defined by equation [4].

$$x_i(\mathbf{I}, \mathbf{W}) = \sum_{j=0}^{n} w_{ji}\, in_j \qquad [3]$$

$$out_i[x_i(\mathbf{I}, \mathbf{W})] = \frac{1}{1 + \exp[-x_i(\mathbf{I}, \mathbf{W})]} \qquad [4]$$

In equation [4] (transfer function) the term $out_i[x(\mathbf{I},\mathbf{W})]$ denotes the output of the processing element PE_i. The transfer function of the processing element PE_i determines the element's output, $out_i[x(\mathbf{I},\mathbf{W})]$, based on its input values $\mathbf{I} = (in_0, in_1, in_2, \ldots, in_j, \ldots, in_n)$ and the corresponding weights, $\mathbf{W} = (w_{0i}, w_{1i}, w_{2i}, \ldots, w_{ji}, \ldots, w_{ni})$, via the *combination function* $x_i(\mathbf{I},\mathbf{W})$ (see equation [3]). The indices $1, 2, \ldots, j, \ldots, n$ indicate the processing elements or inputs that are connected to PE_i. The special *bias input*, in_0, is always fixed to +1, and together with the weight w_{0i} it is used to represent a threshold. Learning in a neural network is the process of setting (adapting) the best weights on the inputs of each processing element. The goal is to produce weights where the output of the network is as close to the desired output as possible for as many examples in the training set as possible. The most common algorithm for doing this is the (error) backpropagation algorithm [Dayhoff, 1996]. The standard implementation of this algorithm proceeds as follows:

1. The network is presented with a training sample or case, defined by the pair (C,*class label*), and, based on the existing weights, calculates and predicts the outputs (e.g., classification) for the case in question.
2. The backpropagation algorithm then determines the error by taking the difference between the calculated outputs and the expected outputs (as provided by a teacher).
3. The error is then fed back through the network and the weights are adjusted (according to some scheme) to minimize the error.
4. Learning normally proceeds until a certain termination criterion is reached. A typical criterion is the *prediction accuracy threshold*, which is often specified somewhere between 80% and 90%. If too low a value is chosen, the network may not be able to generalize well, so the expected performance on unseen test data will be low. Too high a value entices the net into memorizing or *overfitting* the training data. As a consequence the net will not perform well on unseen test data since it has failed to generalize from the training data.

Despite the high number of variables (ca. 7,000 genes) and the low number of observations (n=72), we demonstrated that the standard backpropagation algorithm is a useful tool for analyzing microarray data [Dubitzky et al., 2000]. We carried out three experiments with learning accuracy thresholds of 80%, 85%, and 90% respectively over a 6-fold cross-validation process. The average classification performances were as follows: 93.94%, 94.63%, and 89.70% respectively. Clearly, the decreased performance with the learning threshold of 90% is due to the well-known overfitting effect.

SUMMARY AND CONCLUSION

We presented an overview of two widely used machine learning algorithms and two complementary techniques (cross-validation and multi-model approaches) that have been demonstrated to be useful for microarray analysis. The major advantage of these techniques is that they can be readily applied to explore microarray data. In a recent study [Dubitzky et al., 2000] we showed that decision trees (C5.0 implementation with 20-fold boosting) outperformed neural networks (standard backpropagation algorithm). The added advantage of decision trees is that their results can be directly inspected by humans. This may potentially lead to a deeper understanding of the mechanisms governing the underlying biological entities and process. With standard neural networks this option is not readily available. Perhaps one of the most important drawbacks of the presented methods is that they do not readily lend themselves to single-event probabilities or statistical confidence intervals in connection with their predictions. Clearly, much more analysis, experiments, and statistical and biological validation is necessary to establish a comprehensive set of properties on data mining and machine learning approaches to microarray analysis.

REFERENCES

Arning, A., Agrawal, R., Raghavan, P., "A linear method for deviation detection in large databases", *Proc. 2nd Int'l Conference on Knowledge Discovery and Data Mining*, pp164-169, 1996.

Berry, M.J.A. and Linoff, G., "Data Mining Techniques: For Marketing, Sales, and Customer Support", John Wiley & Sons, 1997.

Breiman, L., Friedman, J.H., Olshen, R.A., Stone, C.J., "Classification and Regression Trees", Wadsworth, Monterey, CA, 1984.

Chen J.J., Wu R., Yang P.C., Huang J.Y., Sher Y.P., Han M.H., Kao W.C., Lee P.J., Chiu T.F., Chang F., Chu Y.W., Wu C.W., Peck K., "Profiling expression patterns and isolating differentially expressed genes by cDNA microarray system with colorimetry detection", *Genomics*, 1;51(3), pp313-24, 1998.

Dayhoff, J.E., "Neural Network Architectures: An Introduction", Thomson Computer Press, 1996.

Dubitzky, W., Granzow, M., Berrar, D., Bulashevska, S., Conrad, C., Gerlich, D., Eils, R., "A Comparison of Symbolic and Subsymbolic Machine Learning Approaches to Molecular Classification of Cancer and Gene Identification", in Critical Assessment of Techniques for Microarray Data Mining (CAMDA-2000*)*, pp12-13, 2000. (extended abstract)

Dubitzky, W., Krebs, O., Eils, R., "Minding, OLAPing, and Mining Biological Data: Towards a Data Warehousing Concept in Biology", *Proc. Network Tools and Applications in Biology (NETTAB), CORBA and XML: Towards a Bioinformatics Integrated Network Environment*, pp78-82, Genoa, Italy, 2001.

Golub T.R., Slonim D.K., Tamayo P., Huard C., Gaasenbeek M., Mesirov J.P., Coller H., Loh M.L., Downing J.R., Caligiuri M.A., Bloomfield C.D., Lander E.S., "Molecular classification of cancer: class discovery and class prediction by gene expression monitoring", *Science* 286(5439) pp531-537, 1999.

Granzow M., Berrar D., Dubitzky W., Schuster A., Azuaje F. and Eils R., "Tumor Classification by Gene Expression Profiling: Comparison and Validation of Five Clustering Methods ", in *ACM-SIGBIO Letters*, 2001. (in press)

Hipp J., Güntzer U., Nakhaeizadeh G., Algorithms for Association Rule Mining: A General Survey and Comparison, *SIGKDD Explorations*, 2(1), pp58-64, 2000.

Mitchell, T.M., "Machine Learning", McGraw-Hill, Singapore, 1997.

Newton M.A., Kendziorski C.M., Richmond C.S., Blattner F.R., Tsui K.W., "On differential variability of expression ratios: improving statistical inference about gene expression changes from microarray data", *J Comput Biol.*, 8(1), pp37-52, 2001

Ng, K., "An experimental comparison of stacking, bagging, boosting for combining ensembles of radial basis function classifiers", Term Project 9.520, Spoken Language Systems Group, Laboratory for Computer Science, Massachusetts Institute of Technology, 1996.

Quinlan, J.R. and Quinlan, R., "C4.5: Programs for Machine Learning", Morgan Kaufmann Series in Machine Learning, 1997.

Upal, M.A. and Neufeld, E., "Comparison of unsupervised classifiers", in *Proc. of the First International Conference on Information, Statistics and Induction in Science*, pp342-353, World Scientific, Singapore, 1996.

Voit E.O. & Radivoyevitch T., "Biochemical systems analysis of genome-wide expression data", *Bioinformatics*, 16(11), pp1023-37, 2000.

Witten, I.H. and Frank, E., "Data Mining: Practical Machine Learning Tools and Techniques with Java Implementations", Morgan Kaufmann Pub., 2000.

2

EVOLUTIONARY COMPUTATION IN MICROARRAY DATA ANALYSIS

Jason H. Moore and Joel S. Parker
Program in Human Genetics, Department of Molecular Physiology and Biophysics, Vanderbilt University Medical School, Nashville, TN 37232-0700

Abstract: We are facing an information explosion in the biomedical sciences. For example, our ability to measure the expression levels of thousands of different genes simultaneously in a particular cell or tissue has far outpaced our ability to store, manage, and analyse the data being generated. In this review, we explore the use of evolutionary computation for dealing with some of the difficult statistical and computational challenges that have resulted from the development and implementation of new technologies such as DNA microarrays. We review genetic algorithms and genetic programming as evolutionary computation strategies that have been applied to the analysis of DNA microarray data.

Key words: Evolutionary computation, genetic algorithms, genetic programming, machine learning, DNA microarrays, gene expression

INTRODUCTION

Our ability to measure the relative expression levels of thousands of genes simultaneously in complex biological systems presents some very difficult statistical and computational challenges. One of these challenges lies in the selection of gene expression variable subsets for statistical evaluation. For example, with 10,000 variables there are approximately $5 * 10^7$ possible subsets of size two, $1.7 * 10^{11}$ possible subsets of size three, and $4.2* 10^{14}$ possible subsets of size four. Clearly, the combinatorial magnitude of variable selection precludes an exhaustive search of all possible variable subsets. One solution to this problem is to first select a subset of variables that each has a statistically significant independent main or marginal effect as evaluated by a univariate statistic such as the t-test or the z-score test of

Thomas et al. [2001]. Although this approach may reduce the size of the search space, it has the potential of missing gene expression variables whose effects are solely or partially through interactions. This is a well-characterized problem in evolutionary biology [Templeton, 2000] and genetic epidemiology [Ritchie et al., 2001]. Since the only way to identify gene-gene interactions is to evaluate variable subsets, and since it is not computationally feasible to evaluate all combinations, machine learning search approaches [Langley, 1996] must be employed. The goal of this review is to introduce evolutionary computation (EC) as a machine learning search strategy for selecting gene expression variable subsets and optimising statistical modelling.

TRAVERSING RUGGED FITNESS LANDSCAPES

One goal of microarray data analysis is to identify a set of expressed genes along with a statistical model that is predictive of some biological or clinical endpoint. The ability to identify the best combination of genes and the best statistical model will largely depend on the total number of possible gene combinations and statistical models as well as the degree to which similar models predict. Collectively, this can be referred to as the fitness landscape where fitness is defined as how well a particular gene expression model predicts the endpoint. Fitness landscapes have been very helpful for understanding biological evolution by natural selection [Kauffman, 1993]. Figure 1 illustrates two hypothetical fitness landscapes. The first represents a smooth fitness landscape in which models that differ only slightly with respect to the combination of genes or statistical functions have similar fitnesses. Thus, if we could list all the possible models along the x-axis with their respective fitness on the y-axis, we would see a Mt. Fuji-like landscape. Finding the best model in this type of landscape is relatively simple and is analogous to a blindfolded mountain climber. Starting at the base of Mt. Fuji, the mountain climber takes steps in different directions and continues climbing in the direction that is sensed to be up. When steps in all directions lead down, the climber removes the blind-fold and realizes that he/she has found the highest peak.

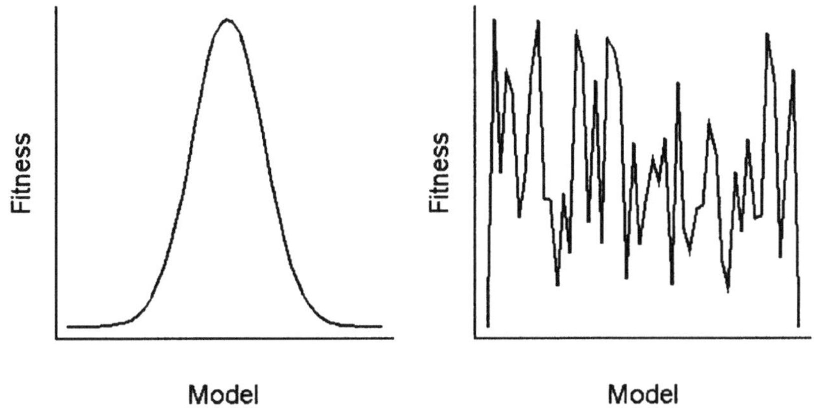

Figure 1. Hypothetical examples of a smooth Mt. Fuji fitness landscape (left) and a rugged Bryce Canyon fitness landscape (right).

The second part of Figure 1 illustrates a rugged landscape that might resemble Bryce Canyon, for example. Here, models that differ only slightly with respect to the combination of genes and/or statistical functions can have very different fitnesses. A blindfolded climber in this Bryce Canyon-like landscape would have a very difficult time finding the highest peak. Rugged fitness landscapes are likely to be very common in complex biological systems that are driven by interactions among system components. In fact, this idea is not new. Wright [1932] emphasized that the relationship between genes and biological endpoints is dependent on dynamic interactive networks of genes and environmental factors. This idea holds true today. Gibson [1996] stresses that gene-gene and gene-environment interactions must be ubiquitous given the complexities of intermolecular interactions that are necessary to regulate gene expression and the hierarchical complexity of metabolic networks.

Assuming rugged fitness landscapes will be the norm rather than the exception, how will we identify optimal combinations of genes and statistical models that are predictive of biological and clinical endpoints? Computer science has provided several strategies for searching fitness landscapes. The first type of search is referred to as an exhaustive or blind search and proceeds by trying every possibility. This is a very successful strategy when the size of the search space is small and every possibility can be enumerated. However, with hundreds or thousands of gene expression variables the search space is effectively infinite. Hill climbing is a second general class of algorithms. With these algorithms, a single model is selected from the search space and evaluated. This model is changed in some way and, if better than the previous model, is retained. If worse, it is discarded and the original model changed again. This is analogous to the blindfolded mountain climber taking steps in different directions and proceeding in the direction that is sensed to be up. Hill climbing algorithms work very well in smooth Mt. Fuji-like fitness landscapes and have the advantage of requiring little computer memory since only one model is evaluated at a time and all others are discarded. An example hill climbing algorithm is simulated-annealing [Kirkpatrick et al., 1983].

An alternative to exhaustive and hill climbing algorithms is the beam search [Langley, 1996]. Beam searches operate on many models simultaneously using an evaluation function to evaluate the fitness of each model. Here, the best solutions are retained and all of the others discarded. The remaining models are then changed and revaluated. This process is iterated many times. Beam searches have the advantage of being able to search many parts of the fitness landscape simultaneously (i.e. in parallel) and are analogous to having hundreds or thousands of blindfolded mountain climbers in Bryce Canyon looking for the highest peak. However, this is at the cost of computer memory because many models need to be evaluated and retained. Genetic algorithms (GAs) and genetic programming (GP) are examples of search methods that utilize a beam search [Langley, 1996; Banzhaff, 1998]. We discuss these forms of evolutionary computation in more detail in the next sections.

GENETIC ALGORITHMS

Evolutionary computation (EC) had its beginnings as early as the 1950s with several independent groups addressing whether evolution could be used for optimisation. That is, can we get computers to solve problems in the same manner that biological populations undergoing evolution by natural selection do? Genetic algorithms (GAs) are perhaps the best-described and best-developed method of EC [Holland, 1975; Goldberg, 1989; Mitchell, 1996].

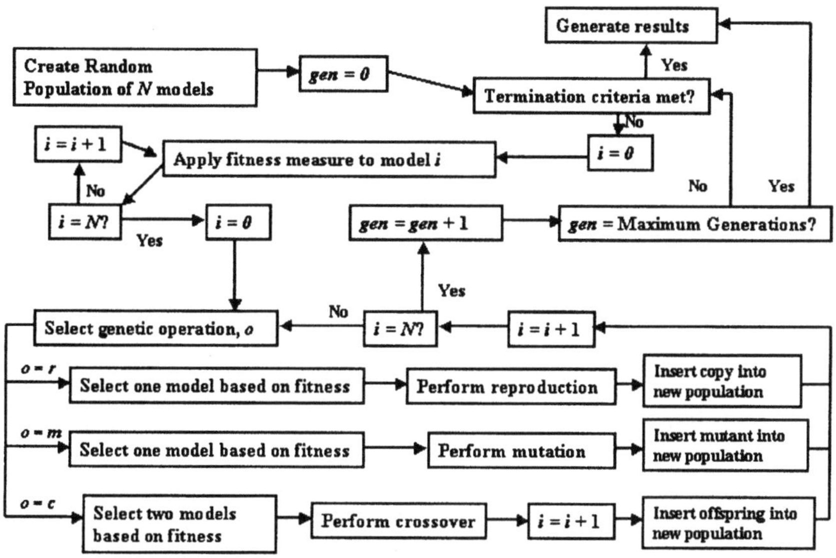

Figure 2. A general flowchart for genetic algorithms and genetic programming (adapted from Koza et al., 1999).

In general, GAs carry out a beam search with changes to models in the beam or population occurring through the use of operators that are inspired by biological mechanisms for generating DNA sequence variability. Figure 2 illustrates a general flowchart for GAs. The first step is to randomly generate a population of models that are represented by binary arrays, the fundamental data structure for the GA. Deciding how to encode models as binary arrays can sometimes be a challenge, depending on the model complexity. The GA population size needs to be large enough and diverse enough to provide the basic model building blocks that will be needed for the beam search yet not so large that it cannot find the right combination of building blocks to converge [Haupt and Haupt, 1998]. The next step is to evaluate the fitness of each model in the population. For microarray data, the fitness might be an estimate of the classification and/or prediction error of a particular combination of gene expression variables and a particular statistical model. The goal is to identify a gene expression model that minimizes the classification and/or prediction error. Once a fitness value has been assigned to each model, the next step is reproduction. Here, the best models are selected either deterministically or probabilistically and model variability is introduced using operators such as recombination and mutation.

With recombination, two good models are selected and pieces of each model are exchanged at some random point or points along the binary array. The general idea is that good models will contain good model building blocks. Recombination (also referred to as crossover in the GA literature) serves to generate different combinations of good model building blocks with the goal of generating even better models. Recombination is the key to the success of GAs just as it is the key to introducing variability among chromosomes during meiosis in biological populations undergoing evolution by natural selection. Other operators such as mutation, inversion, and even transposition have been implemented in GAs; however, they are not as effective as recombination.

Generation 0

Chromosomes	x	f(x)
00011010	26	676
10010000	144	20736
00011100	28	784
00001001	9	81

Generation 1

Chromosomes	x	f(x)
10010000	144	20736
10011100	158	24964
00010000	16	256
00011100	28	784

Figure 3. Example of the first two generations of a simple genetic algorithm maximizing a mathematical function. Note that selection and recombination of the two highest fit solutions results in a new solution with an even higher fitness.

Figure 3 illustrates a very simple example with a population of four models in which the binary arrays encode integer values for the variable X where the goal is to maximize the function $f(X) = X^2$. Thus, $f(X)$ is the fitness function and X^2 is the fitness to be maximized. In generation 0, the four binary arrays are generated using a random number generator. Each is decoded into the corresponding integer value and the fitness evaluated. Then, two models with the highest fitnesses are selected (deterministically in this case) and recombined at a random place. The two new models and the two original highly fit models are retained in the population and passed to the next generation where they are revaluated. The models with the worst fitness are not retained. This process is iterated until some fitness criterion is met, until the population converges to one highly fit solution, or until some computational time limit has expired. Note that one of the two new models generated by recombination has a higher fitness than either of the parents.

Genetic algorithms have been shown to be much more effective at searching Bryce Canyon-like fitness landscapes than either exhaustive or hill climbing algorithms [Goldberg, 1989; Haupt and Haupt, 1998; Mitchell, 1996] and have been shown to be effective at selecting subsets of variables [Yang and Hanover, 1998]. In genetics and molecular biology, GAs have been applied to designing oligonucleotides for hybridisation [Kel et al., 1998], multiple molecular sequence alignment [Zhang and Wong, 1997], protein folding [Dandekar and Argos, 1992], construction of phylogenetic trees [Reijmers et al., 1999], haplotyping in pedigrees [Tapadar et al., 2000], mapping quantitative trait loci [Carlborg et al., 2000], and genetic epidemiology of complex diseases [Congdon et al., 1993]. More recently, GAs have been applied to optimising the modelling of DNA microarray data. We review this work below.

GENETIC PROGRAMMING

With GAs, the models or problem solutions must be encoded by a binary array. However, this may not always be possible. Koza [1992] developed genetic programming (GP) as a flexible way to evolve computer programs to solve problems. Thus, any problem solution that can be encoded by a computer program can be the focus of EC. For example, Koza et al. [1999] demonstrated that GP could evolve simulated electrical circuits that could perform various tasks such as temperature sensing. Genetic programming follows the same basic flowchart illustrated in Figure 2 (described in detail above). Figure 4 illustrates the same function optimisation problem described above except implemented using GP. Here, the solutions are encoded as binary expression trees with arithmetic functions providing even more flexibility for generating integer values than the binary arrays used with the GA. Binary expression trees are a commonly used type of data structure that provides a great deal of flexibility for statistical and mathematical modelling.

Although GP has not been around as long as GAs, the flexibility of the approach has resulted in numerous theoretical and applied studies. Koza et al. [1999] estimate that between 1987 and 1998 there were approximately 1,382 papers published on GP. In genetics and molecular biology, GP has been applied to characterizing protein interactions [Handley, 1994; Raymer et al., 1996], recognizing patterns in protein and DNA sequences [Handley, 1993, 1996; Koza, 1994; Koza and Andre, 1996], and reverse engineering of metabolic pathways [Koza et al., 2001]. Two recent applications of GP to DNA microarray data analysis are reviewed below.

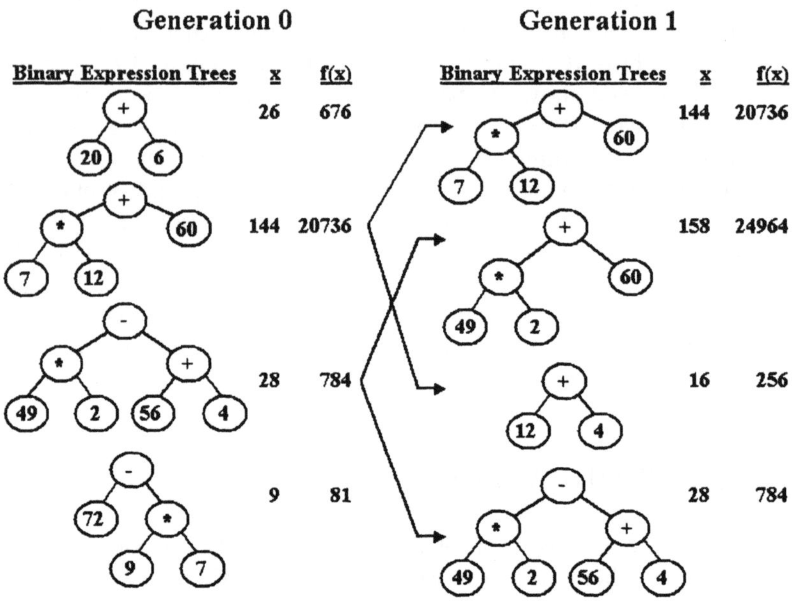

Figure 4. Example of the first two generations of a simple genetic program maximizing a mathematical function. Note that selection and recombination of the two highest fit solutions results in a new solution with an even higher fitness.

PARALLEL IMPLEMENTATION

Both GAs and GP have been shown to be very effective at searching rugged Bryce Canyon-like fitness landscapes. Still, although the beam search that both employ reduces the possibility of finding a locally optimum peak, GAs and GP are not immune to this problem. This motivated the development of parallel or distributed GAs [Cantu-Paz, 2000] and GP [Andre and Koza, 1996]. Here, the GA/GP population is divided into subpopulations called demes that are partially isolated with periodic migration of highly fit models from each subpopulation to all or some of the others. This prevents individual populations from converging on a locally optimum peak because new highly fit individuals are periodically arriving to increase the population diversity. In biology, it is believed that evolution progresses faster in semi-isolated demes than in a single population of equal size [Wright, 1943]. Indeed, there is some evidence that parallel GAs actually converge to a solution much faster than serial or single-population

GAs [Cantu-Paz, 2000]. This superlinear speedup may be due to additional selection pressure from choosing migrants based on fitness [Cantu-Paz, 2000]. When the search space is effectively infinite and the fitness landscape is rugged, parallel GAs and GP may provide better search results.

APPLICATIONS OF EVOLUTIONARY COMPUTATION TO MICROARRAY DATA ANALAYSIS

Clustering methods have emerged as a popular approach to DNA microarray data analysis because they are able to consider the full vector of gene expression variables to perform class discovery. Further, most of the clustering methods such as k-means clustering can be implemented on a desktop computer using commercially available statistical software packages such as S-Plus and SAS and some freely available packages such as Cluster [Eisen et al., 1998]. However, these methods have some important limitations. One of the more important limitations is that they do not always find optimum clusters. This is because methods such as agglomerative and divisive hierarchical clustering use a greedy algorithm that locks observations into a particular cluster that is deemed best at that point in the algorithm but that may not be best globally when all information is considered.

In response to this limitation, Lozano and Larranaga [1999] and Cowgill et al. [1999] developed clustering approaches that are optimised using GAs. Both groups demonstrated using simulated and real datasets that the GA-optimised clustering algorithms were an improvement over other proposed methods. However, these approaches have yet to be applied to microarray data. Recently, Li et al. [2001] (see Chapter 6 of this volume) developed a GA-optimised k-nearest neighbours (KNN) approach to classification. Here, the GA is used to select combinations of genes that are then evaluated using a standard KNN algorithm [Parzen, 1962]. Thus, the GA chromosome or individual consists of a binary array where each linear segment of the array encodes an integer that specifies the index number of a gene from the list of all genes. The fitness function for this GA is the number of observations that are correctly classified by the KNN algorithm. Li et al. [2001] applied this GA-KNN algorithm to the leukemia microarray data of Golub et al. [1999]. The authors first determined the optimal number of genes to include on the fixed length GA chromosome. This length was determined to be 40 genes. The authors then ran the GA 20,000 times and selected a subset of 50 genes that appeared most frequently. These 50 genes were then evaluated for their predictive ability using the KNN algorithm in an independent dataset. A prediction error of approximately 0.03 (33 out of 34) was obtained. This study illustrates the use of GAs to select gene expression variable subsets for analysis using a traditional unsupervised approach.

An alternative approach using GP to select gene expression variables and build classifier rules was developed by Gilbert et al. [2000]. Here, the building blocks of the model were gene expression variables, arithmetic functions (+, -, *, /), and a Boolean 'if greater than or equal to' function that returns a 1 if the first argument is greater than or equal to the second, and 0 otherwise. These building blocks were used to construct linear classifier rules of the form "if $(X_1 * X_2) \geq (X_3 - X_4)$ then group=1, else group=0" where the *X*s are gene expression variables, for example. Gilbert et al. [2000] applied this GP approach to classifying genes as belonging to one of six functional groups using the yeast microarray data of Eisen et al. [1998]. A five-subpopulation parallel GP was run 2,000 generations with migration of the top 5% of models to each population every 10 generations. Using this approach, the authors were able to classify and predict group membership with a high degree of accuracy and even discovered some patterns not previously found in the literature. This study demonstrates the usefulness of GP for deriving classifier rules.

Moore et al. [2001] have developed a symbolic discriminant analysis (SDA) approach to mining gene expression patterns that uses GP to select gene expression variable subsets and to build symbolic discriminant functions. Their goal was to develop a flexible discriminant analysis approach that didn't have the same linear model restrictions as Fisher's linear discriminant analysis [Fisher, 1936]. With SDA, model building blocks are gene expression variables and mathematical functions of any type. In this initial study, the authors used the four basic arithmetic functions (+, -, *, /). Models were represented by binary expression trees similar to those illustrated in Figure 4. The authors applied this method to the human leukemia dataset of Golub et al. [1999]. A sample of 38 acute myeloid leukemia (AML) and acute lymphoblastic leukemia (ALL) tumours were used as a training set to identify optimal SDA models. Here, 10-fold cross-validation [Ripley, 1996] was used to estimate the classification error and prediction error of the models. Models with the minimum prediction error were then evaluated for their predictive ability in an independent sample of 34 AML and ALL tumours. This approach identified two models that correctly classified 38/38 samples in the training dataset and correctly predicted 33/34 samples (prediction error = 0.03) in the independent test dataset. This study illustrates the usefulness of GP for optimising the selection of gene expression variables and symbolic discriminant functions.

SUMMARY

It is increasingly clear that the computational and statistical challenges associated with DNA microarray data analysis are not trivial. Two important challenges include the selection of gene expression variable subsets from among an effectively infinite number of possible combinations and the selection of an appropriate mathematical or statistical model. Evolutionary computation approaches such as GAs and GP use evolution by natural selection as a model for searching rugged fitness landscapes. These approaches have been applied to a variety of problems in genetics and molecular biology and are now being applied to optimisation of microarray data analysis. Studies by Li et al. [2001], Gilbert et al. [2000], and Moore et al. [2001] illustrate the usefulness of GAs and GP for automatically selecting subsets of gene expression variables and for optimising statistical methods for class discovery and class prediction. We anticipate that the flexibility and robustness of these search algorithms will continue to have a positive impact on microarray data analysis.

ACKNOWLEDGEMENTS

This work was supported by the Vanderbilt-Ingram Cancer Center and NIH grants DK58749 and CA90949.

NOTES

Inquiries on obtaining symbolic discriminant analysis (SDA) software should be directed to Dr. Jason H. Moore (moore@phg.mc.vanderbilt.edu). Inquiries on obtaining the software described by Li et al. [2001] and Gilbert et al. [2000] should be directed to the corresponding authors for those studies. General information about obtaining evolutionary computation software can be found at the CMU Artificial Intelligence Repository located at http://www.cs.cmu.edu/afs/cs/project/ai-repository/ai/areas/genetic/0.html.

REFERENCES

Andre D, Koza JR (1996) A parallel implementation of genetic programming that achieves superlinear performance. In Arabnia HR (ed) Proceedings of the International Conference on Parallel and Distributed Processing Techniques and Applications, Volume III, pp 1163-1174. CSREA, Athens.

Banzhaff W, Mordin P, Keller RE, Francone FD (1998) Genetic Programming: An Introduction. Morgan Kaufmann Publishers, San Francisco.
Cantu-Paz (2000) Efficient and Accurate Parallel Genetic Algorithms. Kluwer Academic Publishers, Boston.
Carlborg O, Andersson L, Kinghorn B (2000) The use of a genetic algorithm for simultaneous mapping of multiple interacting quantitative trait loci. Genetics 155:2003-2010.
Congdon CB, Sing CF, Reilly SL (1993) Genetic algorithms for identifying combinations of genes and other risk factors associated with coronary artery disease. In: Proceedings of the Workshop on Artificial Intelligence and the Genome. Chambery.
Cowgill MC, Harvey RJ, Watson LT (1999) A genetic algorithm approach to cluster analysis. Comput Math App 37:99-108.
Dandekar T, Argos P (1992) Potential of genetic algorithms in protein folding and protein engineering simulations. Protein Eng 5:637-645.
Eisen MB, Spellman PT, Brown PO, Botstein D (1998) Cluster analysis and display of genome-wide expression patterns. Proc Natl Acad Sci USA 95:14863-8
Fisher RA (1936) The use of multiple measurements in taxonomic problems. Ann Eugen 7:179-188.
Gibson G (1996) Epistasis and pleiotropy as natural properties of transcriptional regulation. Theor Popul Biol 49:58-89.
Gilbert RJ, Rowland JJ, Kell DB (2000) Genomic computing: explanatory modeling for functional genomics. In Whitley D, Goldberg D, Cantu-Paz E, Spector L, Parmee I, Beyer H-G (eds) Proceedings of the Genetic and Evolutionary Computation Conference, GECCO-2000, pp 551-557. Morgan Kauffman Publishers, San Francisco.
Goldberg DE (1989) Genetic Algorithms in Search, Optimization, and Machine Learning. Addison-Wesley, Reading.
Golub TR, Slonim DK, Tamayo P, Huard C, Gaasenbeek M, Mesirov JP, Coller H, Loh ML, Downing JR, Caligiuri MA, Bloomfield CD, Lander ES (1999) Molecular classification of cancer: class discovery and class prediction by gene expression monitoring. Science 286:531-537.
Handley S (1993) Automatic learning of a detector for alpha-helices in protein sequences via genetic programming. In Forrest S (ed) Proceedings of the 5th International Conference on Genetic Algorithms, ICGA-93, pages 271-278. Morgan Kaufman Publishers, San Francisco.
Handley S (1994) The prediction of the degree of exposure to solvent of amino acid residues via genetic programming. Proc Int Conf Intell Syst Mol Biol 2:156-160.
Handley S (1996) A new class of function sets for solving sequence problems. In Koza JR, Goldberg DE, Fogel DB, Riolo RL (eds) Genetic Programming 1996: Proceedings of the First Annual Conference, pages 301-308. MIT Press, Cambridge.
Haupt RL, Haupt SE (1998) Practical Genetic Algorithms. John Wiley & Sons, Inc., New York.
Holland JH (1975) Adaptation in Natural and Artificial Systems. University of Michigan Press, Ann Arbor.
Kauffman SA (1993) The Origins of Order. Oxford University Press, New York.
Kel A, Ptitsyn A, Babenko V, Meier-Ewert S, Lehrach H (1998) A genetic algorithm for designing gene family-specific oligonucleotides sets used for hybridization: the G protein-coupled receptor protein superfamily. Bioinformatics 14:259-270.
Kirkpatrick S, Gelatt C, Vecchi M (1983) Optimization by simulated annealing. Science 220:671-680.

Koza JR (1992) Genetic Programming: On the Programming of Computers by Means of Natural Selection. MIT Press, Cambridge.

Koza JR (1994) Evolution of a computer program for classifying protein segments as transmembrane domains using genetic programming. Proc Int Conf Intell Syst Mol Biol 2:244-252.

Koza JR, Andre D (1996) A case study where biology inspired a solution to a computer science problem. Pac Symp Biocomput 500-511.

Koza JR, Bennett III FH, Andre D, Keane MA (1999) Genetic Programming III: Darwinian Invention and Problem Solving. Morgan Kauffman Publishers, San Francisco.

Koza JR, Mydlowec W, Lanza G, Yu J, Keane MA (2001) Reverse engineering of metabolic pathways from observed data using genetic programming. Pac Symp Biocomput 434-445.

Langley P (1996) Elements of Machine Learning. Morgan Kaufmann Publishers, Inc., San Francisco.

Li L, Bushel P, Pedersen L, Darden T, Hamadeh H, Bennett L, Afshari C, Paules R, Umbach D and Weinberg C. Computational analysis of leukemia microarray expression data using the GA/KNN method and other existing tools. In press.

Lozano JA, Larranaga (1999) Applying genetic algorithms to search for the best hierarchical clustering of a dataset. Pattern Recog Lett 20:911-918.

Mitchell M (1996) An Introduction to Genetic Algorithms. MIT Press, Cambridge.

Moore JH, Parker JS and Hahn LW (2001) Symbolic discriminant analysis for mining gene expression patterns. Lecture Notes in Artificial Intelligence, in press.

Parzen E (1962) On the estimation of a probability density function and mode. Ann Math Stat 33:1065-1076.

Raymer ML, Punch WF, Goodman ED, Kuhn LA (1996) Genetic programming for improved data mining: An application to the biochemistry of protein interactions. In Koza JR, Goldberg DE, Fogel DB, Riolo RL (eds) Genetic Programming 1996: Proceedings of the First Annual Conference, pages 375-380. MIT Press, Cambridge.

Reijmers TH, Wehrens R, Buydens MC (1999) Quality criteria of genetic algorithms for construction of phylogenetic trees. J Comput Chem 20:867-876.

Ripley BD (1996) Pattern Recognition and Neural Networks. Cambridge University Press, Cambridge.

Ritchie MD, Hahn LW, Roodi N, Bailey LR, Dupont WD, Plummer WD, Parl FF and Moore JH (2001) Multifactor Dimensionality Reduction Reveals High-Order Interactions among Estrogen Metabolism Genes in Sporadic Breast Cancer. Am J Hum Genet 69:138-147.

Tapadar P, Ghosh S, Majumder PP (2000) Haplotyping in pedigrees via a genetic algorithm. Hum Hered 50:43-56.

Templeton AR (2000) Epistasis and Complex Traits. In: Wade M, Brodie III B, Wolf J (eds) Epistasis and Evolutionary Process. Oxford University Press.

Thomas JG, Olson JM, Tapscott SJ, Zhao LP (2001) An efficient and robust statistical modeling approach to discover differentially expressed genes using genomic expression profiles. Genome Res 11:1227-1236.

Wright S (1932) The roles of mutation, inbreeding, crossbreeding, and selection in evolution. Proceedings of the 6th International Congress of Genetics 1:356-366.

Wright S (1943) Isolation by distance. Genetics 28:114-138.

Yang J, Honavar V (1998) Feature subset selection using a genetic algorithm. IEEE Intelligent Systems 13:44-49.

Zhang C, Wong AK (1997) A genetic algorithm for multiple molecular sequence alignment. Comput Appl Biosci 13:565-581.

3

USING NON-PARAMETRIC METHODS IN THE CONTEXT OF MULTIPLE TESTING TO DETERMINE DIFFERENTIALLY EXPRESSED GENES

Gregory Grant[1], Elisabetta Manduchi, Christian Stoeckert, Jr.
Center for Bioinformatics, University of Pennsylvania (PCBI)

Abstract: Our focus is on the Golub et al. ALL/AML oligo-nucleotide array data set [Golub et al., 1999] with regard to the question of determining differentially expressed genes between pairs of sample types. We use this data set to analyze methods of determining genes which are likely to be differentially expressed between ALL T-cells and ALL B-cells. To this end, we employ non-parametric methods, in the context of multiple testing, for attaching statistical measures of confidence to genes predicted to be differentially expressed. In particular, we apply the method of using t-statistics, with p-values calculated through permutations, and with the Westfall and Young step-down approach to correct for multiple testing, developed by Dudoit et al. [Dudoit et al., 2000]. We also use PaGE [Manduchi et al., 2000], developed at PCBI, for assigning confidence to predictions by calculating false-positive rates directly from empirical "gene-independent" distributions. We compare the performance of these methods on the Golub et al. data. We exploit the large sample size to analyze the effect of the number of observations on a variety of issues relating to the prediction of differential expression, in particular to the reproducibility of results. In addition, we investigate the concept of using shifted intensities for data such as the Golub et al. data set. We also investigate the usage of "absent calls" in oligo-nucleotide array data.

Key words: non-parametric methods, multiple testing, differential expression, t-statistics, Westfall and Young step-down, PaGE, confidence measures, biological replicates, false-discovery rate, clinical heterogeneity

[1] Contact: ggrant@pcbi.upenn.edu

INTRODUCTION

In Golub et al. [Golub et al., 1999], Affymetrix data [Lockhart et al., 1996] have been generated for leukemic myeloid and lymphoblastic cells. The data contain a relatively large number of observations of the expression levels of several cell types. In particular the data contain many biological replicates; that is, the observations were taken from different cells and different individuals. The data also represent a heterogeneous collection of cell types. As such, this data set presents an opportunity to study the effect of the sample size on the results obtained by the current methods which are aimed at giving the least conservative measures of significance for predictions of differential expression. We use the term "prediction" here because array data give only indications of which genes are likely to be differentially expressed. These predictions may then be corroborated by other experimental approaches.

Much of the classical theory that might be used to tackle this problem depends on assumptions about the forms of the distributions of the gene intensities. In particular, standard techniques such as the two-sample *t*-test, depend on these distributions being normal (see [Mood et al., 1963] for example). The Golub et al. data give us many observations, each observation being one array experiment. Examining the gene intensity distributions over the observations shows that many of them are not normal. Some have apparently uniform distributions, some unimodal, others bimodal, or even trimodal. Moreover, the inclusion of absent calls introduces further irregularity into the distributions. Such heterogeneity precludes the possibility of transforming the distributions into standard distributions and the use of non-parametric methods becomes necessary. By non-parametric methods we mean those that make minimal assumptions about the form of the distributions involved. In highly parallel gene expression data, the challenge of having irregular distributions is compounded by issues of multiple testing. Non-parametric approaches for the problem of assigning confidence measures to the prediction of differential expression, in the context of multiple testing, have been proposed using standard and non-standard methods [Dudoit et al., 2000; Manduchi et al., 2000]. These methods take as input multiple observations for each of two (or more) sample types. We use the Golub et al. data as a pool and sample from it, each time applying these algorithms and tabulating the results. This gives information about how reproducibility depends on the sample size.

We also address a complementary issue. One might hope to be able to maintain a low false-positive rate when using only a few observations. However, it is key, regardless of which method is used to make predictions, that the data represent sufficiently the variability of the sample types.

Variability is generally of two types: experimental, and biological. With a general definition of sample type, such as "Acute Lymphoblastic Leukemia" (ALL) or "Acute Myeloid Leukemia" (AML), biological variability will be the overwhelming factor between the two. AML cells, for example, can be divided into many subcategories, such as M1, M2, etc., male/female, peripheral blood/bone marrow, etc. ALL cells can be either T-cells or B-cells, and each of these divides into respective subcategories. It is also expected, since they are cancerous cells, that they will fall into unknown subcategories of tumor types, perhaps related to the success/failure response to treatment. If the sample type is represented by a few experiments only, then the probability of them falling into one or more subcategories becomes almost a certainty. The Golub et al. data give us a chance to see the effect of this clinical heterogeneity on predictions of differential expression, as a function of sample size. Instead of focusing on the ALL and AML classification, we controlled for the heterogeneity of the sample type as much as possible by focusing on the B-cell and the T-cell experiments within the ALL group. This gives us two reasonably homogeneous sample types, for which we still have many observations.

TYPES OF DIFFERENTIAL EXPRESSION

Generally for two sample types A and B of interest, most genes will not be differentially expressed between them (see [Sagerstrom et al., 1997]). No gene however is expressed at exactly the same level in a given sample type every time it is measured. Instead each gene will have a distribution of intensities, with respect to observations of the sample type. A sample type can be general, such as "hematopoietic cells," or specific, such as "B-cell lymphoblasts" at a particular stage of development, under particular conditions. The distributions will depend heavily on this definition of sample type.

Genes will be differentially expressed in a non-deterministic manner, meaning that the two distributions overlap to some degree, though they still have different means. A deterministically differentially expressed gene, if it existed, would never be incorrectly predicted, regardless of the sample size for each sample type. Figure 1 shows an example of a gene that appears to be nearly deterministically differentially expressed between ALL B-cells and ALL T-cells. This is the exception, however. Between ALL B-cells and ALL T-cells, almost all genes which are differentially expressed appear to be non-deterministic. Figure 2 shows such a gene.

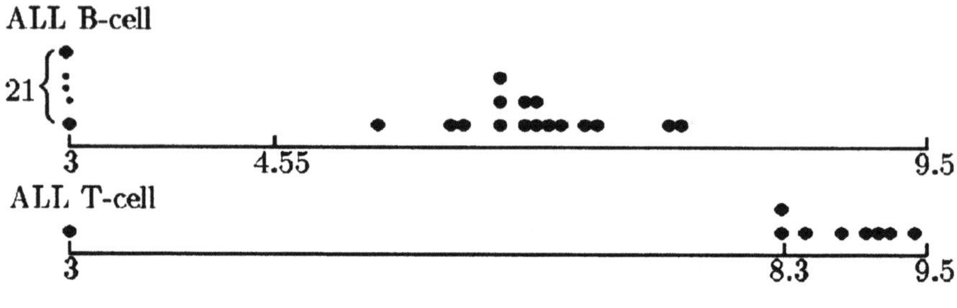

Figure 1. GenBank Accession: U23852, T-lymphocyte specific protein tyrosine kinase. Data points are taken from the Golub et al. data set. The top plot is of the expression in the 37 ALL B-cells used, 21 of which were absent calls, indicated on the left. The lower plot is of the expression of the same tag in the 9 ALL T-cells, with one absent call. Graphs are in *ln* scale.

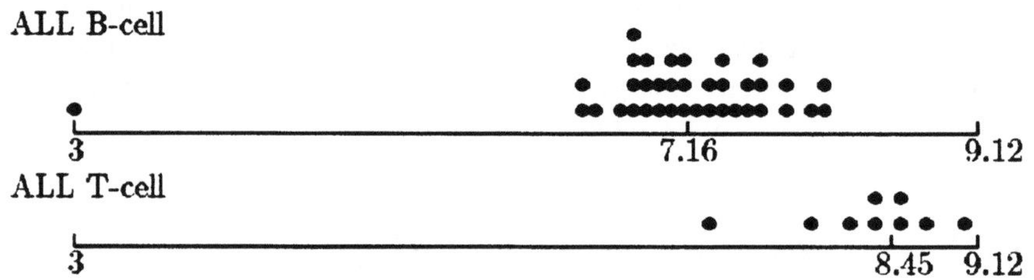

Figure 2. GenBank Accession: M23323 T-CELL SURFACE GLYCOPROTEIN CD3 EPSILON CHAIN PRECURSOR. Data points are taken from the Golub et al. data set. Notation is as in Figure 1. Gene displays non-deterministic differential expression.

Absent Calls

When performing statistical analyses of oligo-nucleotide data, one must decide how to handle the absent calls. Absent calls present a complication, in that if we set them to a minimal value, then they tend to create highly bimodal distributions, such as in Figure 3, where the "29" indicates that there are 29 absent calls, which were set to the minimal intensity (3 on this graph, which is in *ln* scale). The decision to include absent calls was based on a comparison of expected results when we both do and do not include them. In the T-CELL ANTIGEN CD7 PRECURSOR ID:M37271 (see Figure 3), there is negligible difference in the means of the intensities of the present calls between the B-cells and the T-cells. When absent calls are included, however, the means separate clearly (4.04 versus 8.03).

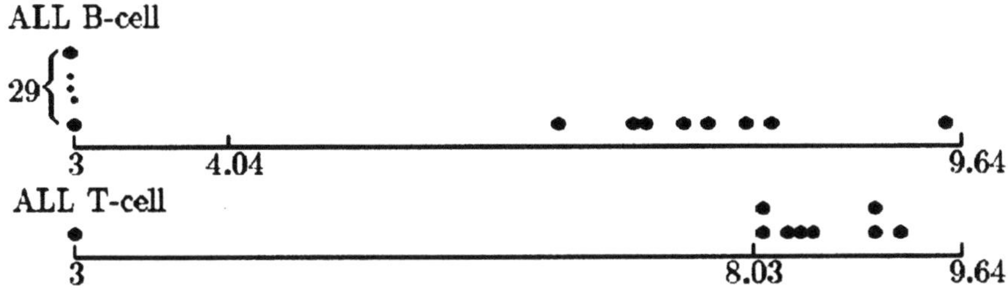

Figure 3. GenBank Accession: D00749, T-CELL ANTIGEN CD7 PRECURSOR. Data points are taken from the Golub et al. data set. Notation is as in Figure 1. The twenty nine intensities at level 3 are absent calls, (set to the minimum intensity $\ln(20) \approx 3$). The mean of the ALL B-cell distribution with the absent calls included is 4.04. The distribution for the same gene, over the nine ALL T-cell experiments, has one absent call and mean 8.03.

Among the B-cell experiments, there are roughly 80% absent calls for this gene, whereas among the T-cell experiments there are roughly 10%. It is only the absent calls that are differentiating this gene between B- and T-cells. This phenomena was consistent across many T-cell specific genes in the data set. Based on this, absent calls were included for all genes that had at least one non-absent call intensity, by setting them to the minimum value of 20. Genes that had absent calls in all experiments, for both cell types, were eliminated from further analysis, leaving approximately 5000 genes.

A consequence of the inclusion of absent calls is that it dramatically increases the non-deterministic nature of the differential expression, as a high percentage of genes were absent in one or more experiments. As a result, some genes which gain inordinately large variance may not be detected after the inclusion of absent calls. We compared predictions with and without absent calls. With PaGE [Manduchi et al., 2000], one T-cell specific gene (see Figure 4) was picked up when absent calls were not included that was not picked up when they were included. In this case, the absent calls did not sufficiently separate the means to overcome the large variance present in the expression of this gene. On the other hand, five known differentially expressed genes were detected when absent calls were included, that were not seen without absent calls, including the one in Figure 3.

Some T-cell genes were not detectable by any methods we used. Figure 5 shows such a gene. In this case the differential expression is too subtle to detect, given that approximately 5000 simultaneous tests are performed. If this were the only gene under investigation, then the available number of observations might be sufficient to detect it. However, when looking at thousands of genes simultaneously, this kind of separation is likely to occur just by chance for many genes that are not differentially expressed.

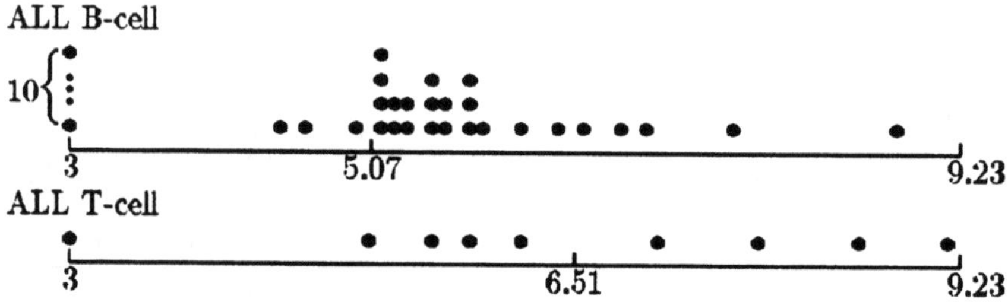

Figure 4. GenBank Accession: M30894, TCRG T cell receptor gamma chain. Data points are from the Golub et al. data set. Notation is as in Figure 1.

The less specific is the definition of the sample type, the more biological variability we expect to find for more genes. Correspondingly, the less specific is the definition of the sample type, the greater percentage of the differentially expressed genes will have highly overlapping distributions. These issues should be considered when designing array experiments to search for differentially expressed genes. The more non-deterministic is the gene expression, the more observations will be necessary to make accurate predictions.

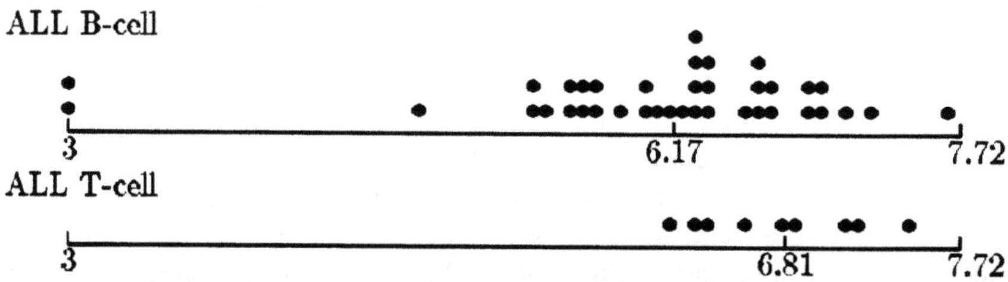

Figure 5. GenBank Accession: X94232, Novel T-cell activation protein. Data points are taken from the Golub et al. data set. Notation is as in Figure 1. Differential expression of this gene is too subtle to detect in the context of the thousands of genes on the array.

There are two general approaches to controlling the reliability of the predictions. The classical approach is to control the *experiment-wise* Type I error. In this case the probability of making any false predictions at all is controlled (see [Ewens et al., 2001] Section 12.4).

An alternative approach is to control the false-discovery rate. In this case if predictions of upregulation are being made for type A versus type B, and if M predictions are made with V of them being incorrect, then the false-discovery rate is the expected value of V/M, denoted by $E(V/M)$.

Successfully controlling the experiment-wise Type I error instead of the false-discovery rate gives a set of high confidence results, but at the possible cost of a higher false-negative rate. For example it might be preferable to tolerate 5-10% of false positives among the predictions if it means finding many more true positives.

The two methods we investigate below take these two different approaches. We start with PaGE which controls the false-discovery rate.

THE PaGE APPROACH

PaGE (Patterns from Gene Expression) [Manduchi et al., 2000] is part of a larger software package developed at PCBI for analyzing gene expression data. One of its features is that it reports estimated confidence measures on predictions of differential expression. The confidence measure reported is one minus the false-discovery rate. PaGE essentially controls for the false-positive rate and then basically translates that into the false-discovery rate using Bayes theorem.

Given any criteria for predicting differential expression, the false-positive rate of this criteria is the percentage of the set of non-differentially expressed genes consisting of genes which satisfies the criteria. When trying to find a small set of differentially expressed genes from a large pool, a seemingly reasonable false-positive rate, say .05, can lead to a set of predictions with a high proportion of false positives among them. For example, if 50 out of 1000 genes are differentially expressed, a false-positive rate of .05 will give 950(.05) = 47.5 false positives, on average. There can be at most 50 true positives, so if there are 47.5 false positives there will be at best 50/(50 + 47.5)% = 51.28% confidence in any given prediction, and that will be achieved only if the false-negative rate is zero.

For each gene g, let $X_{g,A}$ and $X_{g,B}$ be the random variables giving the intensities of gene g when measured in a sample of types A and B, respectively. Let $\overline{X}_{g,A}$ (respectively $\overline{X}_{g,B}$) be the average of n_A (respectively n_B) random variables each having the distribution of $X_{g,A}$ (respectively $X_{g,B}$). For a given gene g, we say that it is upregulated at group B as compared to group A if

$$\frac{\mu_{g,B}}{\mu_{g,A}} > 1$$

where $\mu_{g,A}$ (respectively $\mu_{g,B}$) denotes the (unknown) true mean of $X_{g,A}$ (respectively $X_{g,B}$). A brief summary of the PaGE approach follows:

Let s% be the desired false-positive rate. We want to determine C (the "upper cut-ratio") such that, if we say that a gene g is up-regulated at group B as compared to group A when

$$\frac{\overline{x}_{g,B}}{\overline{x}_{g,A}} > C$$

then the false-positive rate is expected to be no greater than s%, where $\overline{x}_{g,A}$ (respectively $\overline{x}_{g,B}$) are the observed values of $\overline{X}_{g,A}$ (respectively $\overline{X}_{g,B}$).

The false-positive rate is the probability that, if a gene is chosen at random from the set of genes which are true negatives, then the ratio of its sample means between the two groups, $\frac{\overline{X}_{g,B}}{\overline{X}_{g,A}}$, is greater than C. Since the set of genes which are true negatives is the set of genes for which $\mu_{g,B} / \mu_{g,A} \leq 1$, this probability is bounded above by the probability that, if a gene is chosen at random from the set of genes which are true negatives, then

$$\frac{\dfrac{\overline{X}_{g,B}}{\mu_{g,B}}}{\dfrac{\overline{X}_{g,A}}{\mu_{g,A}}} > C. \qquad [1]$$

Since only a small percentage of genes are expected to be true positives, we approximate this probability by the (unconditional) probability that [1] holds.

Key ingredient: we approximate the unknown distribution of $\frac{\overline{X}_{g,A}}{\mu_{g,A}}$ by that of

$$\frac{\frac{X_{g,A}}{\overline{X}_{g,A}}-1}{\sqrt{n_A-1}}+1 \qquad [2]$$

Similarly for group *B*. In [Manduchi et al., 2000] we show that this approximation holds in the case of normally distributed gene intensities that are sufficiently bounded away from zero. Generalizations of this result to other distributions have been verified by simulation (see [Manduchi et al., 2000]). Simulations were also conducted to justify using the approximation on the Golub et al. data. Further empirical tests of this heuristic will appear elsewhere.

We use [2] to approximate the distribution of $\frac{\overline{X}_{g,B}/\mu_{g,B}}{\overline{X}_{g,A}/\mu_{g,A}}$, with the empirical distribution from the data. The random quantities in this expression are g, $\overline{X}_{g,A}$, and $\overline{X}_{g,B}$. This allows us to use [1] to calculate a C by numerical integration of this empirical distribution. This gives us the desired false-positive rate as a function of C.

Approximating [2] with empirical distributions from the data, C is solved for in [1] by numerical integration.

Finally, the confidence in the predictions is related to the false-positive rate via Bayes Theorem.

For down-regulation, we proceed in a similar fashion to determine c (the "lower cut-ratio") such that, if we say that a gene g is down-regulated at group B as compared to group A when

$$\frac{\overline{x}_{g,B}}{\overline{x}_{g,A}} < c,$$

then the false-positive rate is expected to be no greater than s%

The theory above gives no indication of the power of the test. We are investigating this issue using a simulated data study, to appear elsewhere.

We now turn to a more classical approach, but which uses a different measure of confidence, an *experiment-wise* measure.

THE *t*-STATISTIC STEP-DOWN APPROACH

The other method employed has been developed by Dudoit et al. [Dudoit et al., 2000]. This method controls the experiment-wise Type I error. It uses *t*-statistics, with *p*-values calculated through permutations, and with the Westfall and Young [Westfall et al., 1993] stepdown approach to correct for multiple testing. We used the one-sided version of the method, to test specifically for up-regulation in ALL T-cells versus ALL B-cells (software to implement this can be downloaded at http://www.cbil.upenn.edu/tpWY). In estimating the null hypothesis distribution for the t-statistic, entire experiments (columns) are permuted, so as to preserve the dependencies between the genes. This method achieves a chosen experiment-wise Type I error. This means that if the desired Type I error is set at .05, then there is a 95% chance of *no* false positives. This is in contrast to PaGE, which achieves a desired confidence in the set of predictions. So that if PaGE reports a confidence of 95%, then 95% of the predictions should be true positives.

We now apply these two methods to repeated samplings from the Golub et al. data.

REPLICATE STUDY RESULTS

We first used the analyses of [Dudoit et al., 2000] to find genes which are up-regulated in ALL T-cells versus ALL B-cells. We used 37 B-cell experiments[2] and the nine T-cell experiments from the Golub et al. data set. For various values of *n* we conducted the analysis 100 times, each time choosing a random sampling of *n* of the B-cell experiments, and testing them against all nine of the T-cell experiments. This was done so as to control for the effect of variation from all but one source, that being the number of B-cell experiments. In practice, variability will come from both sources, so that the true sample size needed to obtain reliable predictions may be even greater than our results indicate. The fraction of times a gene was predicted as up-regulated among the 100 experiments was recorded. A portion of the results is given in Table 1. The full table has 106 rows.

A similar table has been generated using PaGE. A portion of the results is given in Table 2. The complete table contains 81 rows.

It is apparent from these tables that there are not many genes which will be predictably differentially expressed without many observations. Notice

[2] Experiment 72 was accidentally omitted by one of our parsers, so this experiment ended up being eliminated in the study.

that even the gene shown in Figure 1 is not reliably detected using the step-down approach, regardless of the sample size.

Table 1. Column labelled *n* gives the fraction of times the gene was predicted by the methods of [Dudoit et al., 2000] as up-regulated in ALL T-cells versus ALL B-cells, out of 100 comparisons between *n* randomly chosen B-cells and all 9 T-cell experiments, from the Golub et al. data set. Intensities were pre-processed by taking natural logs. The (experiment-wise) Type I error was taken to be 0.1.

36	25	20	15	10	9	8	7	6	5	4	Gene Information
1	1	1	1	1	1	.99	.6	0	0	0	CD3G CD3G antigen, gamma polypeptide
1	1	1	1	.77	.64	.6	.71	0	0	0	T-CELL DIFFERENTIATION ANTIGEN
1	1	1	.96	.69	.61	.45	.4	.22	.29	0	PTPN7 Protein tyrosine phosphatase
1	1	.79	.73	.67	.75	.48	.62	.52	.02	0	T-CELL ANTIGEN CD7 PRECURSOR
1	1	1	.96	.78	.58	.4	.27	.38	0	0	TCF7 Tcell specific Trans. factor 7
1	.92	.79	.59	.31	.26	.14	.11	.04	.08	0	SMT3A protein
1	.98	.9	.6	.11	.08	.04	0	0	.01	.03	Chromosome 17q21 mRNA clone LF113
1	.62	.41	.3	.09	.11	.03	.02	.03	0	.06	Trophinin mRNA
.01	.35	.44	.46	.34	.34	.1	.09	0	.01	0	T-CELL SURFACE GLYCOPROTEIN CD3
0	.06	.13	.24	.2	.21	.15	.22	.18	.08	0	T-CELL ANTIGEN CD7 PRECURSOR
0	.07	.17	.25	.21	.14	.13	.15	.06	0	0	MPHASE INDUCER PHOSPHATASE 2
.32	.15	.09	.05	.01	.03	.03	0	.02	.01	.03	GB DEF = Splicing factor, ...
0	0	.02	.06	.05	.11	.03	.05	.11	.13	0	Protein tyrosine kinase related ...
0	.14	.19	.1	.02	.03	0	.01	.01	.01	.01	MACHalpha2 protein
0	0	.04	.14	.08	.07	.09	.08	.01	0	0	ANX1 Annexin I (lipocortin I)
0	0	.03	.03	.07	.09	.05	.03	.04	.04	0	CD47 CD47 antigen (Rhrelated ...
0	0	0	.03	.08	.04	.02	.04	.07	.08	0	KIAA0050 gene
0	.04	.07	.05	.06	.07	0	.02	.01	.04	0	RNA-BINDING PROTEIN FUS/TLS
0	.06	.05	.08	.03	.03	.02	.01	0	.02	.01	DNA-BINDING PROTEIN NEFA PRECURSOR
0	0	.04	.06	.08	.06	.05	.01	0	0	0	Lactate dehydrogenase B gene ...
0	0	0	.04	.05	.07	.08	.03	0	0	0	MXS1 Membrane component, X ...
0	.05	.07	.03	.01	.03	.01	.01	0	.01	.03	Lipidactivated protein kinase
0	0	0	.02	.05	.04	.05	.09	0	0	0	HIG1 mRNA
0	0	.01	.06	.02	.05	.03	.01	.01	.04	0	GB DEF = T-lymphocyte specific ...
0	0	0	.02	.06	.05	.03	.04	0	0	0	Signal transducer and activation ...
0	0	.01	.05	.02	.05	.02	.01	.02	.02	0	GB DEF = Karyopherin beta 3 mRNA
0	0	0	.03	.03	.04	.02	.02	.04	0	0	GATA3 GATAbinding protein 3
0	.01	.02	.07	.05	.01	.02	0	0	0	0	TXN Thioredoxin
0	0	0	.01	.02	.05	.04	.02	.04	0	0	TYROSINEPROTEIN KINASE ITK/TSK
0	.03	.04	.01	.02	0	.01	0	.01	0	.04	GB DEF = Inducible nitric oxide ...
0	0	.01	.02	.03	.07	0	0	.02	.01	0	GB DEF = T-cell antigen receptor ...

This is because with this many observations the differential expression is not strong enough to have been statistically significant every time, in the context of the other approximately 5000 genes being analyzed simultaneously. This gene was, however, picked up by virtually all PaGE runs with more than two

B-cell observations. The table shows how the dependability of the results drops dramatically as the sample size is reduced.

Table 2. Column labelled n gives the fraction of times the gene was predicted by PaGE as up-regulated in ALL T-cells versus ALL B-cells, out of 100 comparisons between n randomly chosen B-cells and all 9 T-cell experiments, from the Golub et al. data set. Confidence was set at 90%, and a shift of 5000 was used (see section SHIFTS).

36	25	20	15	10	9	8	7	6	5	4	3	2	Gene Information
1	1	1	1	1	1	1	1	1	1	1	1	.88	TCRB T-cell receptor, beta cluster
1	1	1	1	1	1	1	1	1	1	1	1	.73	TCRB T-cell receptor, beta cluster
1	1	1	1	1	1	1	1	1 .	1	1	1	.73	GB DEF = MAL gene exon 4
1	1	1	1	1	1	1	1	1	1	1	.99	.64	GB DEF=T-lymphocyte specific protein ...
1	1	1	1	1	1	1	1	1	1	1	.95	.57	GB DEF=T-cell antigen receptor gene ...
1	1	1	1	1	1	1	1	1	1	1	.93	.43	Protein tyrosine phosphatase PTPCAAX2...
1	1	1	1	1	1	1	1	1	1	1	.92	.4	CHIT1 Chitinase 1
1	1	1	1	1	1	1	1	1	1	1	.95	.35	Na,KATPase gamma subunit mRNA
1	1	1	1	1	1	1	1	1	1	.97	.88	.38	NATURAL KILLER CELLS PROTEIN 4 PRECURSOR
1	1	1	1	1	.99	.96	.93	.89	.87	.95	.81	.43	T-CELL ANTIGEN CD7 PRECURSOR
1	1	1	1	1	1	1	1	.98	.98	.88	.53	.15	GB DEF = Selenoprotein W (selW) mRNA
1	1	1	1	1	1	1	1	1	1	.93	.48	.1	CD1B CD1b antigen (thymocyte antigen)
1	1	1	1	1	1	.99	.99	.95	.89	.84	.64	.18	ADA Adenosine deaminase
1	1	1	1	1	1	1	1	1	.98	.87	.38	.09	TCF7 T-cell specific Trans. factor 7
1	1	1	1	1	1	1	.99	.97	.93	.82	.48	.11	T-CELL SURFACE GLYCOPROTEIN CD3 ...
1	1	1	.99	.99	.99	.96	.96	.89	.85	.7	.52	.24	MIC2 Antigen identified by monoclonal...
1	1	1	1	.75	.72	.77	.88	.86	.87	.9	.5	.12	TCRD T-cell receptor, delta
1	1	.99	.96	.86	.83	.75	.78	.76	.75	.76	.58	.34	PSAP Sulfated glycoprotein 1
1	1	1	.99	.96	.93	.85	.76	.64	.46	.22	.07	.01	GB
1	.86	.83	.77	.76	.77	.62	.68	.53	.51	.47	.31	.1	mRNA fragment encoding betatubulin. ...
1	.63	.58	.64	.7	.72	.77	.88	.81	.66	.32	.14	.01	T-CELL ANTIGEN CD7 PRECURSOR
1	.71	.7	.72	.59	.58	.51	.53	.55	.48	.39	.26	.14	Macrophage migration inhibitory fact ...
.98	.59	.61	.53	.55	.59	.59	.58	.51	.51	.4	.32	.07	SM22ALPHA HOMOLOG
.07	.29	.4	.47	.6	.5	.51	.54	.62	.51	.45	.32	.07	MPO Myeloperoxidase
.13	.26	.44	.47	.48	.48	.47	.5	.42	.45	.38	.12	.02	Lactate dehydrogenase B gene exon 1 ...
.03	.08	.34	.33	.54	.5	.36	.45	.39	.26	.17	.05	.01	TCRG T- cell receptor gamma chain
0	.11	.32	.32	.36	.36	.3	.43	.35	.35	.2	.07	.01	KIAA0050 gene
0	.03	.13	.22	.29	.33	.27	.42	.41	.37	.27	.24	.07	RPS3 Ribosomal protein S3
0	.06	.12	.22	.3	.4	.35	.41	.39	.28	.24	.11	.05	LTB Lymphotoxinbeta
0	.11	.25	.26	.33	.36	.34	.38	.23	.22	.16	.09	.04	Homolog Drosophila enhancer of split ...
0	.06	.21	.25	.37	.38	.34	.41	.29	.27	.11	.06	.01	CD2 antigen (p50), sheep red blood ...
0	.07	.16	.3	.32	.31	.24	.34	.22	.25	.18	.09	.06	Put. HMG17 protein gene extracted ...
0	.04	.12	.23	.21	.28	.27	.3	.29	.28	.2	.19	.07	Histone H1x
0	.05	.14	.18	.28	.31	.25	.31	.27	.26	.19	.06	.04	H4/g gene for H4 histone
0	.04	.18	.2	.25	.29	.23	.24	.24	.25	.12	.13	.05	PTMA gene extracted from Human ...
0	.04	.14	.19	.31	.23	.21	.33	.26	.29	.13	.07	.01	GLUL Glutamateammonia ligase ...
0	.03	.14	.19	.29	.18	.25	.3	.21	.27	.19	.12	.04	GB DEF=Polyadenylate binding protein ...

It is interesting to compare the results of the two tables. Of the genes that were picked up reliably with many observations (the top 8 in Table 1, and the top 23 in Table 2), only two are in both tables, the "T-CELL ANTIGEN CD7 PRECURSOR" and the "TCF7 Transcription factor 7 (T-cell specific)." Based on this, it is recommended to use both methods, and to combine the results. For the two genes that do appear on both tables, the T-cell antigen is consistently predicted by PaGE with 4 observations, whereas with the step-down method nearly 25 observations were necessary. For the TCF7 Transcription factor, PaGE consistently predicted it with 5 observations, whereas 15 were required for the step-down method.

We take the genes with 1's in the first column (36) to be our most reliable predictions. At the bottom half of Table 2 we see that many genes which have 0's in the first column are predicted as differentially expressed fairly often with fewer observations. This is due to clinical heterogeneity. When the cell type is heterogeneous, such as "ALL B-cells," there are many subclasses that the cell type can be classified into, so that if we choose only a small sample of experiments to represent the class, we are likely to have over-represented some subclasses. We investigate this issue further below where we gauge the sample size necessary to overcome this problem. First, however, it is necessary to investigate the nature of the data at a more fundamental level.

SHIFTS

PaGE predictions of differential expression are based on ratios of mean intensities, as opposed to basing them on differences. The use of ratios allows for the convenient introduction of a "shift" parameter. In particular, we investigate the effect of measuring differential expression by the ratio of the shifted intensities $(x_2 + s)/(x_1 + s)$, where x_1 and x_2 are the respective intensities in two different sample types.

Varying the shift allows for different regions of the data to be emphasized. No shift corresponds to a region between the lines through the origin, as seen in the graph on the left in Figure 6. This corresponds to a criteria such as taking a fixed cut-off, say 2-fold, and predicting anything over 2-fold (or under 1/2-fold) to be differentially expressed. Such a uniform criteria would be justified if the data looked, for example, like that in the scatter plot on the left in Figure 7, where the spread is proportional to the magnitude of expression. A graph of the actual ALL B-cell vs. ALL T-cell data is given in the graph on the right in Figure 7.

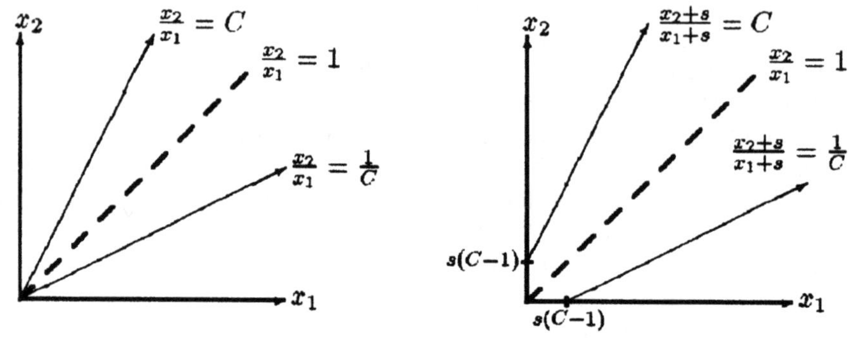

Figure 6. Lines bounding the region of up- and down-regulation expression for the case of no shift (left) and a shift equal to s (right). For simplicity we are assuming the lower cut-ratio c equals the reciprocal of the upper cut-ratio C.

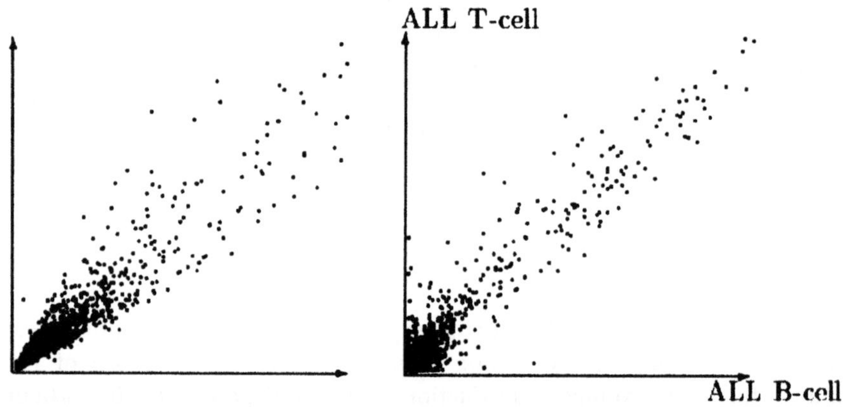

Figure 7. Idealized scatter plot (left) with spread proportional to intensity. Scatter plot of ALL B-cell versus ALL T-cell (right), from the Golub et al. data. Axes represent raw unlogged intensities.

As can be seen from this graph, spread of the data is much greater at the low intensities. This is caused largely by the inclusion of the absent calls. It is therefore clear that a shift is necessary. The shift concept has been used in other methods, e.g. [Newton et al., 2001].

PaGE results were generated by varying the shift, comparing the 9 T-cell to the 37 B-cell experiments used. The results are shown in Table 3. The set of predictions varies as the shift varies, as can also be seen from the graph in Figure 8.

Some genes are predicted as up-regulated in T-cells, regardless of shift, whereas others are predicted for only a range of shifts.

Table 3. Genes predicted by PaGE to be up-regulated in T-cells (9 samples) versus shift B-cells (37 samples). Columns correspond to different shifts of the data. Confidence set to 95%. An asterisk indicates that the gene was predicted as up-regulated in T-cells versus B-cells. A shift of 1000 is considered optimal for this dataset, based on empirical studies of the PaGE estimate [2]. See also Figure 8.

	shift					
ID	200	500	1000	5000	50000	IDinfo
V00599					*	mRNA fragment encoding betatubulin. ...
M13792				*	*	ADA Adenosine deaminase
M16279				*	*	MIC2 Antigen identified by monoclonal antibodies...
M23323			*			TCELL SURFACE GLYCOPROTEIN CD3 EPSILON CHAIN ...
X03934	*	*	*	*	*	GB DEF = Tcell antigen receptor gene T3delta
X00437	*	*	*	*	*	TCRB Tcell receptor, beta cluster
X76223	*	*	*	*	*	GB DEF = MAL gene exon 4
U14603	*	*	*	*	*	Protein tyrosine phosphatase PTPCAAX2 (hPTPCAAX2)...
U23852	*	*	*	*	*	GB DEF = Tlymphocyte specific protein tyrosine ...
M12886	*	*	*	*	*	TCRB Tcell receptor, beta cluster
M59807	*	*	*	*		NATURAL KILLER CELLS PROTEIN 4 PRECURSOR
M28826	*	*	*	*		CD1B CD1b antigen (thymocyte antigen)
U50743	*	*	*	*		Na,KATPase gamma subunit mRNA
M59807	*	*	*	*		NATURAL KILLER CELLS PROTEIN 4 PRECURSOR
U49835	*	*	*	*		CHIT1 Chitinase 1
U67171	*	*	*	*		GB DEF = Selenoprotein W (selW) mRNA
D00749	*	*	*	*		TCELL ANTIGEN CD7 PRECURSOR
M21624	*	*	*			TCRD Tcell receptor, delta
M16336	*	*	*			CD2 CD2 antigen (p50), sheep red blood cell receptor
M37271	*	*	*			TCELL ANTIGEN CD7 PRECURSOR
X59871	*	*	*			TCF7 Transcription factor 7 (Tcell specific)
X14975	*	*	*			GB DEF = CD1 R2 gene for MHCrelated antigen
S78187	*					MPHASE INDUCER PHOSPHATASE 2
L10373	*					MXS1 Membrane component, X chromosome, ...
M28825	*					CD1A CD1a antigen (thymocyte antigen)
X62891	*					Mutant coseg gene for vasopressinneurophysin ...
S56151	*					HMFG
J04132	*					CD3Z CD3Z antigen, zeta polypeptide (TiT3 complex)
L40386	*					DP2 (Humdp2) mRNA
X04145	*					CD3G CD3G antigen, gamma polypeptide (TiT3 complex)
Z19002	*					ZINC FINGER PROTEIN PLZF
L05148	*					Protein tyrosine kinase related mRNA sequence

PaGE relies on the estimation [2], which can become inaccurate as the shift decreases to zero. Empirical tests of this approximation using the Golub et al. data show that a shift of 1000 achieves the best conservative approximation for comparing 37 to 9 experiments having the type of variation present in these data.[3] Even if this PaGE approximation were perfect, a moderate shift would be preferred. This can be seen on the graph in Figure 8.

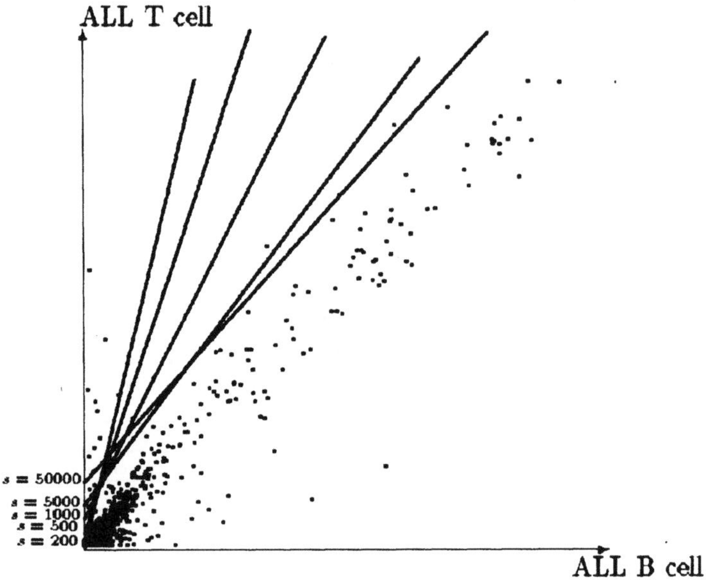

Figure 8. Lines show cut-offs for predictions of up-regulation in ALL T-cells versus ALL B-cells, for varying shifts.

With little or no shift, a very large slope must be used to avoid the false positives at the low intensities. The result is that true positives are missed in every intensity range except the lowest. Likewise, a shift that is too large might tend to pick up false positives from the higher intensity genes. A review of the literature on the 19 genes that were predicted as up-regulated in T-cells, using the shift value of 1000, showed no clear false positives, with most being known to be up-regulated. This gives some empirical verification of the confidence measures in PaGE. On the other hand, of the 10 genes that were predicted to be upregulated with a shift of 200, but were not predicted with higher shifts, there were only four true positives, and five apparent false negatives, with one being unclear. Thus for such a low shift, the confidence appears to have been reduced to around 50%, as might be expected.

[3] In the case of Table 2 we used a uniform shift of 5000, which gave the best approximation, according to the simulations, simultaneously for all values of n used

THE EFFECT OF CLINICAL HETEROGENEITY

If the sample types under question are not sufficiently homogeneous, then using a relatively small sample size gives a high probability that one or more biological subclasses are over represented by the experiments. If this happens, the answers obtained might not be answers to the questions being asked. For example, suppose one is looking for genes that are differentially expressed between B-cells and T-cells, using 10 observations of each type. Consider a gene that is not differentially expressed between B-cells and T-cells, but nonetheless is differentially expressed in differing stages of cell cycle or development. Since there are many genes on the chip, it is not unlikely that the majority of the B-cell experiments are in one state, and the majority of the T-cell experiments in another state, with respect to some of the genes. In this case, we may falsely predict the genes that are differentially expressed between these two states to be differentially expressed between B-cells and T-cells. The statistical methods are correctly predicting differential expression, however between different sample types than expected, which can lead to false conclusions.

To measure the effect of this clinical heterogeneity when using the prediction methods of [Dudoit et al., 2000] and [Manduchi et al., 2000] on data of a similar nature to that of the Golub et al. data, we generated virtual data, based on their ALL B-cell data. To have data that are as variable as theirs, we used the B-cell empirical distributions themselves to generate the virtual data. We sampled randomly 1000 times (with replacement) from the gene tags on the chip. For each of these 1000 gene tags, we sampled from the empirical B-cell distribution of that gene 200 times. This created 200 virtual observations of 1000-gene experiments, containing comparable noise to the actual Golub et al. data. The 1000 genes were sampled independently from each other, so that we have created data with 1000 independent genes. We repeated this making 200 virtual experiments with 3000 independent genes. Certainly the approximately 7000 genes on the Affymetrix chip are not expressed independently in B-cells, however it seems reasonable that there might be 1000 to 3000 independent genes. For each data set, we compared random selections of 5, 10, 20, 30, 40, and 50 experiments against each other. Since all experiments were generated from the same distributions, any predictions of up-regulation are false-positives.

The results are given in Table 4. The effect of clinical heterogeneity is clear when only 5 or 10 observations are used. After 20 observations, the effect effectively disappears, and the percentage of false-positives levels off. This percentage of false-positives is not expected to converge to zero, because the PaGE confidence measure is not 100% (it was 90% in this case).

Table 4. Fraction of times, out of 100 runs, that any predictions of up-regulation were made, comparing a virtual sample to itself, for varying numbers of observations. All such predictions represent false-positives.

	Number of independent genes	
Observations	1000	3000
5	0.39	0.44
10	0.15	0.19
20	0.10	0.11
30	0.06	0.07
40	0.06	0.02
50	0.03	0.04

A similar test was conducted with the step-down method. This method was not susceptible to the problem. This is most likely due to the fact that with only 5 observations, the step-down method, which relies on permutations, makes highly conservative predictions. There are not enough experiments to permute very many times. As this problem diminishes, after roughly 10 observations, the effect of clinical heterogeneity has already diminished. It is interesting to note that in the experiments, the step-down method made false-positive predictions roughly 5-8% of the time, with an experiment-wise Type I error of 0.1. This gives some empirical verification that this reported Type I error is close to correct, and to the extent that they are off, they are conservative.

CONCLUSIONS

We have used the Golub et al. data to investigate the issue of the impact of variability when differential expression predictions are obtained from a small sample size for each sample type. When the desired confidence measures relate to very homogeneous classes of cells, then as few as two or three observations might suffice. However, often heterogeneous sample types are compared, such as "all lymphoblasts" versus "all myeloid cells," to find any discernable differential expression. This study shows that the non-deterministic effect is greater than might be expected. It is necessary to perform further studies characterizing this effect, as many investigators are currently designing array based analyses of their specific systems of interest, and one of the most basic decisions they have to make is how many and what type of experiments to perform. It is not a simple issue, and to address it, it is necessary to have case studies with results that are verified by a large sample. We have used the Golub et al. data for this purpose.

We have also used the data to investigate the performance of two currently existing tools for making predictions of differential expression. At issue is whether to use an experiment-wise Type I error, or the false-discovery rate. The results show that the latter approach allows for greater sensitivity, particularly with few replicates, with the trade-off being the introduction of a small percent of false positives. Each technique predicted some genes that the other technique did not. Therefore it is advisable to use both methods and combine the results.

ACKNOWLEDGEMENTS

We thank Warren Ewens and Sandrine Dudoit for helpful discussions on the step-down method and the false discovery rate, Joan Mazzareli for the literature review to validate the predictions, Jonathan Schug for helpful discussions, and Brian Brunk and Georgi Kostov for help with the E2K cluster. We also thank the rest of the members of the Universtity of Pennsylvania Computational Biology and Informatics Laboratory (CBIL) for their support. This work was supported in part by NIH grants R24-DK56947 and K25-HG-00052-01A1.

REFERENCES

Dudoit S., Yang Y.H., Callow M., Speed T. 2000. "Statistical methods for identifying differentially expressed genes in replicated cDNA microarray experiments." UC Berkeley, Technical report #578. http://www.stat.berkeley.edu/users/terry/zarray/Html/matt.html

Ewens W.J., Grant G.R. 2001. "Statistical Methods in Bioinformatics." *Springer-Verlag* New York.

Golub T.R., Slonim D.K., Tamayo P., Huard C., Gaasenbeek M., Mesirov J.P., Coller H., Loh M.L., Downing J.R., Caligiuri M.A., Bloomfield C.D., Lander E.S. 1999. "Molecular classification of cancer: class discovery and class prediction by gene expression monitoring." *Science* 286(5439): 531-537.

Lockhart D.J. et al. 1996. "Expression monitoring by hybridization to high-density oligonucleotide arrays." *Nature Biotechnology* 14: 1675-1680.

Manduchi E., Grant G.R., McKenzie S.E., Overton G.C., Surrey S., Stoeckert C.J. Jr. 2000. "Generation of patterns from gene expression data by assigning confidence to differentially expressed genes." *Bioinformatics*, 16(8): 685-698. http://www.cbil.upenn.edu/PaGE

Mood A.M., Graybell F.A. 1963. "Introduction to the Theory of Statistics." *McGraw-Hill* New York.

Newton M.A., Kendziorski C.M., Richmond C.S., Blattner F.R., Tsui K.W. 2001 "On differential variability of expression ratios: improving statistical inference about gene expression changes from microarray data." *J. Comput. Biol.*, 8(1):37-52. http://www.biostat.wisc.edu/geda/eba.html

Sagerstrom C.G., Sun B.I., Sive H.L. 1997. "Subtractive cloning: past, present, and future." Ann. Rev. *Biochem.*, 66: 751-783.

Westfall P.H. and Young S.S. 1993. "Resampling-based multiple testing: examples and methods for *p*-value adjustment." Wiley series in probability and mathematical statistics, Wiley.

4

ITERATIVE LINEAR REGRESSSION BY SECTOR

Renormalization of cDNA Microarray Data and Cluster Analysis Weighted by Cross Homology

David B. Finkelstein[1], Rob Ewing[1], Jeremy Gollub[1], Fredrik Sterky[1], Shauna Somerville[1], and J. Michael Cherry[2]

[1]Department of Plant Biology, Carnegie Institution of Washington, Stanford CA,[2]Department of Genetics, Stanford University, Stanford CA

Abstract: Two-color DNA microarray data has proven valuable in high-throughput expression profiling. However microarray expression ratios ($\log_{base2}$ratios) are subject to measurement error from multiple causes. Transcript abundance is expected to be a linear function of signal intensity (y = x) where the typical gene is non-responsive. Once linearity is confirmed, applying the model by fitting log-scale data with simple linear regression reduces the standard deviation of the $\log_{base2}$ratios, after which fewer genes are selected by filtering methods. Comparing the residuals of regression to leverage measures can identify the best candidate genes. Spatial bias in $\log_{base2}$ratio, defined by printing pin and detected by ANOVA, can be another source of measurement error. Independently applying the linear normalization method to the data from each pin can easily eliminate this error. Less easily addressed is the problem of cross-homology which is expected to correlate to cross-hybridization. Pair-wise comparison of genes demonstrate that genes with similar sequences are measured as having similar expression. While this bias cannot be easily eliminated, the effect of this probable cross-hybridization can be minimised in clustering by weighting methods introduced here.

Key words: clustering, normalization, microarray, regression, homology

INTRODUCTION TO THE ITERATIVE METHOD

Empirical observations, validated by statistical tests, indicate that distinct classes of measurement error alter cDNA microarray data. When these measurement errors are detectable and conform to defined models, corrections can be applied during re-normalization. However, supporting biological evidence may be required to validate any normalization method. For the Spellman and Sherlock cell cycle data [Spellman et al., 1998], the spatial and signal intensity dependent measurement errors were corrected through re-normalization. Re-analysis of the Spellman and Sherlock cell cycle data set begins with a new method of normalization that more accurately reduces the effects of outliers and spatial variation on the arrays. First, all background-corrected signal intensity values are log transformed. Then linear regression is performed where the signal intensities of one channel is predicted by the other channel. Spatial error is corrected by performing this regression independently for each sector. Each sector corresponds to a slotted printing pin. The microarrays used in the Spellman experiments had four sectors printed with four distinct pins.

The result is four linear intensity equations, one for each sector. Next residuals are calculated for each of the four regression lines. Outliers (those residuals e where $|e| > 2$ x standard deviation of e) are temporarily removed and the four regression functions are recalculated. If the difference in fit between the new and old regression lines is less than .001, as measured by r-squared, then no further residuals are removed. Else, outliers are removed by the same test as above and the iterations continue. Once completely determined, the slope and intercept values are applied as correction factors to the log transformed channel 2 values. The result is that the function of log channel 1 and log channel 2 closely approximates $y = x$. Then these values are exponentiated, a new ratio is calculated and this ratio is put on the familiar $\log_{base2}$ scale. This re-normalization alone has been demonstrated to substantially reduce the standard deviation of $\log_{base2}$ratios.

Next, the detection of cross-hybridization was examined. Unlike sector bias, which is directly measurable, cross-hybridization is inferred from expression patterns. The yeast genome is fully sequenced, thus the sequences of PCR fragments were known. Therefore it is possible, with some error, to determine the likely number of transcripts that could cross-hybridise to a given PCR fragment. The correlation between the likelihood of cross-hybridization and the frequency of transcripts with cross-homology is difficult to assess without empirical evidence. It is important to note that modelling the molecular events during hybridization has proven difficult. Therefore, no analysis can be used to correct data. However, weighting can be applied as an informed post hoc method. In this way, such analysis may indicate when biological confirmation experiments are warranted, rather than supply a mathematical solution.

APPLYING LINEAR NORMALIZATION

Applying a linear model presumes that the data is or should be linear with respect to intensity. The assumptions that the typical gene is non-responsive and that rare genes are no more likely to respond than abundant genes leads to the adoption of the linear model y = x. The typical gene assumption is probably valid for large-scale genome wide arrays, where only a small number of genes are expected to respond to any given test. For small specialised arrays, where y (dependent variable) is expression under stress condition and x (independent variable) is the expression of genes under the control condition, the typical gene assumption may not be valid. Therefore this equation is valid for the typical case and invalid for the biologically responsive genes. We further presume that biologically responsive genes are sufficiently responsive to be distinguished from the typical gene using statistical tests. For responsive genes of biological interest, the residuals of regression should be especially large.

If intensity functions are non-linear [Roger Bumgarner, MGED3, Stanford, CA], then non-linear lowess regression methods [Yang et al., 2000] may be required instead of the simple linear regression method employed here. In these cases, however, it is still those genes with the highest residuals from the fitted function that are of the greatest interest.

In all tested cases, applying a linear model of error combined with the iterative removal of outlying residuals reduces the standard deviation of the final $\log_{base2}$ratios. The range of the data is not substantially altered. However, the distribution of the data may change. Frequently, the kurtosis increases, meaning that the tails are more widely dispersed with respect to the standard deviation and the skew may change in scale and in direction. Filtering iteratively normalized data without considering spatial bias increased the number of genes that are consistently changed at the $|\log_{base2}\text{ratio}| > 2$ for one of the eleven Elutriation arrays by 4.3% (an increase of 9 genes) when compared to data normalized by the Stanford Microarray Database default method. When the iterative method is applied to each sector to correct spatial problems the number of genes that pass filtering criterion actually decreases. In both cases the overall standard deviation of the data is reduced. Only independent empirical methods can determine whether the differences in analysis methods are removing false positives.

Spatial Methods

Observation based on a spatial display tool developed for microarrays (data not shown) indicated that spatial problems may exist for several Spellman and Sherlock arrays. Re-normalization by sector requires 4 parallel normalizations and assumes that functional groups of genes are not printed together. For many arrays the reduction in standard deviation of the log_{base2}ratios by spatial linear normalization is marginal. However, significant spatial effects have been detected in other cDNA arrays and therefore it is worth testing arrays for the effect.

Spatial bias is detectable with a simple ANOVA (y = log_{base2}ratio and X = grid #) that yields an F-test and r-squared value. Non-parametric methods such as the Kruskal-Wallis test also serve this function [Kruskal and Wallis, 1952). Our current best estimate is that, if r-squared values are below .05, then spatial error is not significant. Best practice may indicate repeating experiments that exhibit substantial sector bias, rather than applying sector specific normalization methods, which are post hoc and may only partially repair the effects.

Applying the Linear Method by Sector

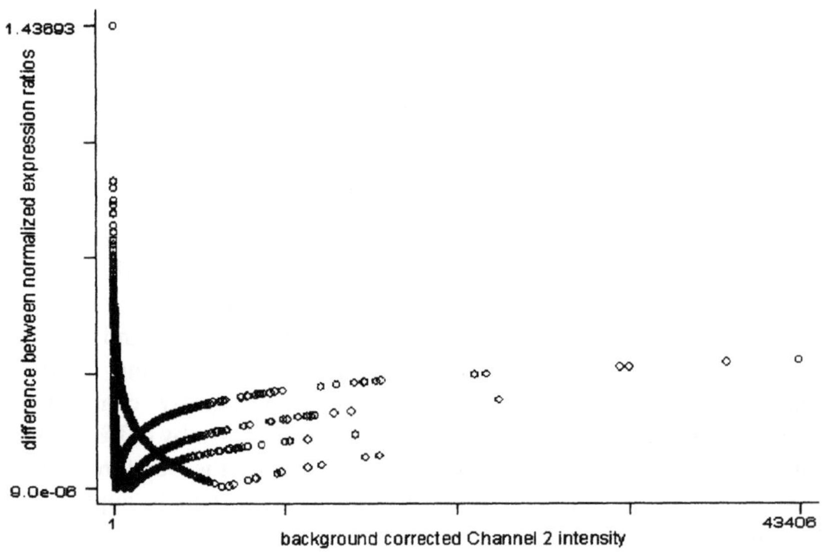

Figure 1. The absolute value of the difference between log_{base2}ratio calculated by the SMD method and the Iterative method is plotted on the y-axis. The background-corrected channel 2 intensity is plotted on the x-axis. Each pattern of spots is derived from a distinct sector.

For each of the four independent sectors of each DNA microarray, the iterative simple linear regression technique is applied. As expected, many arrays are not substantially altered by this approach. However, in instances where outlier sectors are significant, by ANOVA F-tests, differences in normalization are visually evident (Figure 1). Note that the four sectors each have independent patterns with respect to background corrected channel 2 intensity (CH2D). The differences between the SMD method and Iterative method are consistently greater at low intensities (below 150). Each pattern is at a minimum when the linear regression equation for a given sector is equal to the SMD global mean. In the case presented in Figure 2. there is a clear difference in the minimum of one pattern, which may indicate spatial bias in that sector.

RESULTS

Note that the Iterative method consistently reduces the number of genes that pass the filters. It also consistently lowers the standard deviation of the $\log_{base2}$ratios in these studies. It does not, however, consistently improve the global correlation between the $\log_{base2}$ratios of any two arrays (Table 1).

Table 1 Filtering results for two methods. Filtering parameters: all spots that have an average intensity of 100 in each channel and a $|\log_{base2}\text{ratio}|>2$ in at least 1 array were selected.

	SMD Method	Iterative Method	Proportional Difference
α-Factor:	334	269	0.805
Elutriation:	179	135	0.754
CDC:	1204	1099	0.913

Selecting Biologically Significant Genes

Once normalization is completed, significantly differential genes are identified by their expression ratios. Because the $\log_{base2}$ratios are rarely distributed normally, there are often more candidate genes identified than standard deviation-methods would predict. When kurtosis is high, the proportion of differential genes at the tails of the distribution will also be high. Determining which ratios are false positives is critical to reducing cost and effort of confirmation experiments [Tusher et al., 2000].

The analysis of residuals from simple linear regression affords a screen for these outliers. A linear regression function is subject to the influence individual measurements in direct proportion to their distance from the mean. That is, data from genes at extreme intensities have greater leverage (influence) on the predicted regression line than data points near the mean of x. Measures of leverage will select a distinct group of genes from a line

Examples of Changed Arrays

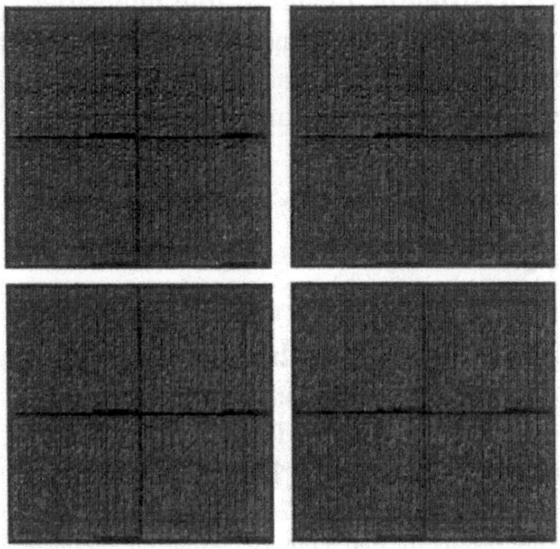

Figure 2. Examples of changes arrays. The plots above show the spatial pattern of log2_ratios on two Elutriation arrays (SMD EXPID 56 (row B) and 57(row A) normalized by the SMD method on the left and by the Iterative method on the right. All spots with a log2_ratio greater than 1 appear in red. All spots with a ratio below 1 appear in green. Black spots indicate a flagged spot, white spots have a ratio of 1. Note that the iterative method (Column 2) partially corrects the spatial bias seen in the data normalized by the SMD method (Column1) for both experiments 56 and 57.

when compared to those found by residuals. Residuals that are distant from the predicted line may have little leverage if they are especially close to the mean. Since most genes are neither especially rare nor abundant, we expect most biological outliers to be near the mean of x (where x is the control state) and therefore have a high residual and a low leverage [Neter et al., 1990]. If we plot leverage versus the square of the residual (Figure 3), we can see a much smaller class of genes are significantly (by standard deviation measures) different by residual measures but not by leverage. Genes that are high in leverage and low in residuals are likely false positives, if they have a differential ratio. Genes that are high in both categories are ambiguous, some can be reasonably expected to be authentically differential. Discerning between these genes is beyond the scope of this simple test.

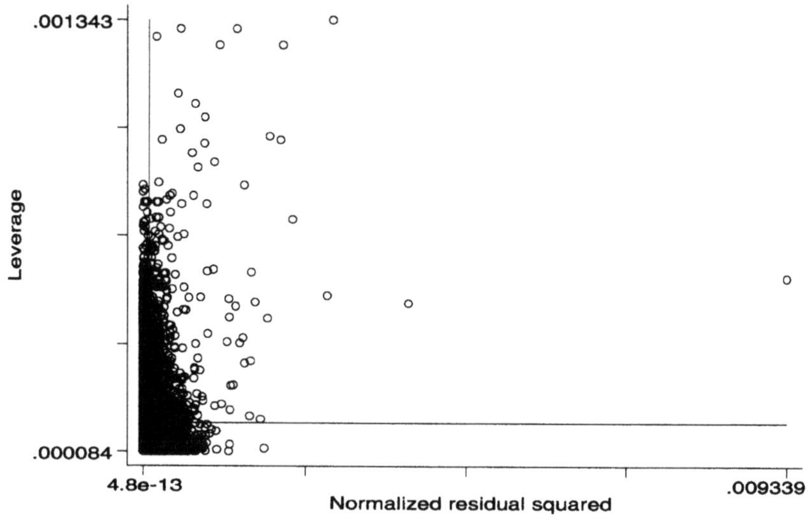

Figure 3. Leverage vs. residual squared plot of microarray data (from lvr2plot procedure in STATA 6.0 Stata Corp, College Station, TX). The lines show the average values of the leverage (y axis) and the normalized residual squared (x axis).

SEQUENCE SIMILARITY IN YEAST ARRAYS

The degree to which cross-hybridization might influence microarray expression data was also examined. First, a preliminary analysis was performed that related sequence similarity to the degree of correlation between expression profiles. Several assumptions are made. First, it was assumed that the full length ORFs available from SGD (Saccharomyces Genome Database) approximate the targets actually used on the microarray. This assumption is deemed reasonable, as yeast primer pairs were designed to include as much of the ORFs as possible [Gavin Sherlock, *personal communication*]. Second, it was assumed that the degree of sequence similarity between a pair of sequences, as measured by an alignment program such as BLASTN, would approximate the degree of cross-hybridization between those sequences.

2,690 ORFS were selected from the original 6,178 yeast ORFs. The selected ORFs were those with the fewest missing expression data values. ORFs with greater than 8 missing values across the 62 experiments were excluded. For all pairs of the 2,690 ORFs, the correlation coefficient between the expression profiles was calculated and a BLASTN alignment of the sequences created. For all pairs of ORFs with some degree of homology, the correlation coefficients were extracted and are plotted as two histograms in Figure 4. ORF pairs are divided according to their BLASTN expect-values. Correlation coefficients for ORF pairs with BLASTN expect-value greater than 1e-4 are shown in grey and those with BLASTN expect-value less than 1e-4 are in black.

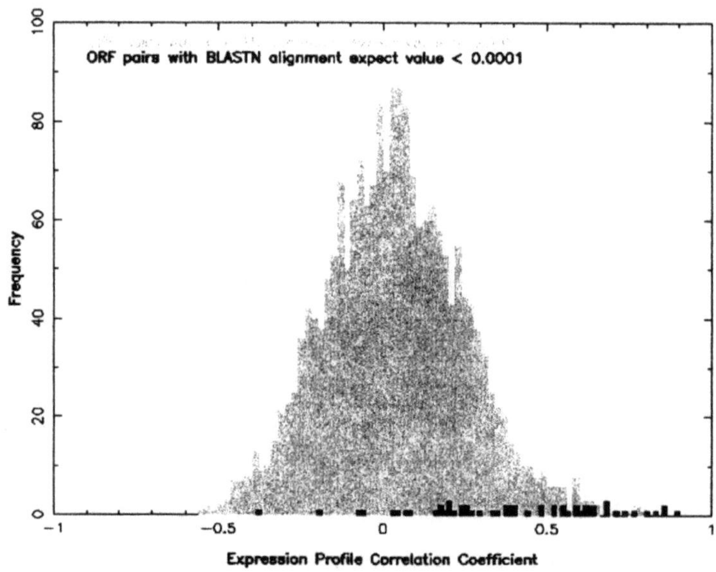

Figure 4. Pairwise correlation coefficients of the expression pattern across 62 experiments of 2,690 yeast ORFS. The histogram in black represents correlation coefficients for ORF pairs with high sequence similarity and the histogram in grey represents correlation coefficients for ORF pairs with no sequence similarity. Note the relative rarity of cross homology and the relatively high degree of co-expression (mainly positive correlation coefficients) amongst the similar ORFs.

Relatively few ORF pairs showed significant sequence similarity. In total, alignments between 2050 ORF pairs were found. Of these alignments, 1991 had expect-values greater than 1e-4 and 59 pairs had expect-values less than 1e-4. The mean expression profile correlation coefficient in the set of 1991 ORF pairs was 0.036, whereas for the set of 59 ORF pairs with more significant expect-values, the mean correlation coefficient was 0.419.

Despite the small numbers, it appears that ORF pairs with a higher degree of sequence similarity are also more likely to exhibit a higher degree of correlation between their expression profiles. The expect-value indicates, but does not prove cross-hybridization. It is also possible that genes with high sequence similarity may have similar function and therefore may be authentically co-expressed. Cross-hybridization and the degree to which this may confound results from genome-wide microarray experiments should definitely be considered in the design of future microarrays and gene-specific probes used wherever possible.

Calculating the Weights for Clustering

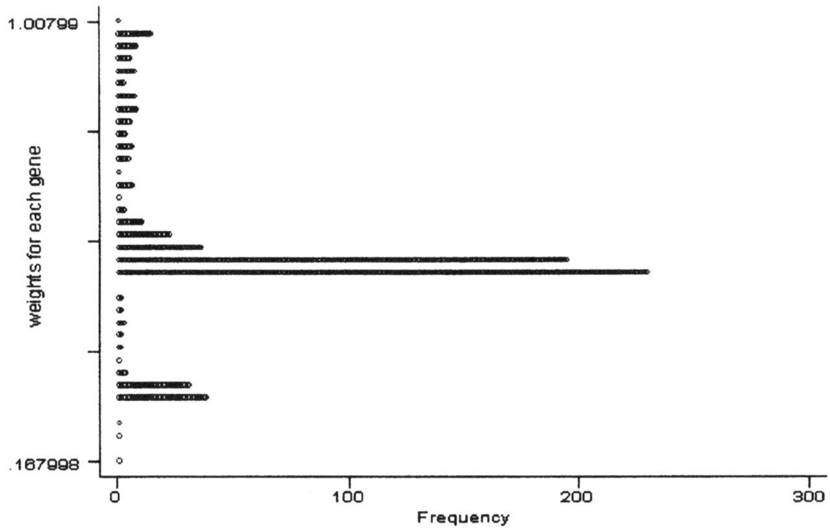

Figure 5. Histogram of the weights applied to potentially cross-hybridizing genes. Note that most weights were nearly 0.5, and only a small sub-population is weighted less than 0.25.

We extended the idea of incorporating measures of sequence similarity into the analysis of microarray data by weighting genes for which cross-hybridization was likely. ORFs were weighted if they had sequence similarity (BLASTN alignment score > 100) with at least one other distinct ORF. 782 alignments met this criterion, representing 678 ORFs. The highest expect-value in this set of alignments was 1e-21. Weights were first calculated for each pair of ORFs in the set. In cases where ORFs matched more than one other ORF, weights were multiplied. Weights were calculated for each ORF pair as the sum: 0.5 + 0.5 (e_m/e_p) where e_m is the lowest expect-value exponent in the set (1e-21) and e_p the expect-value exponent for that ORF pair. For example, for an ORF pair whose alignment had an expect-value of 1e-42, the weight would be calculated as 0.5 + 0.5(-21/-42) = 0.75. In the set of resulting weights, the strongest weight was 0.16 and the weakest weight 1. The distribution of these weights is plotted in Figure 5. Weighting the data according to these weights and then redoing the hierarchical clustering produced relatively little change in the clusters. This is probably due to the relatively low level of cross-homology in this particular data set, and the method would be expected to have greater effect in a data set with a greater degree of cross-homology. Although sequence homology is likely to be a reasonable approximation of cross-hybridization, empirical evidence of cross-hybridization would be preferable.

DETERMINING BEST PRACTICE

Best practice of microarray data analysis is directly tied to the application of the data. If the arrays are to be used as rapid screening tools, then sophisticated normalization and analysis may not be necessary. If, however, the object of the experiment is to model subtle biological patterns across gradients of time or treatment, much more complex analysis is required. Furthermore, while statistical measures are useful in measuring and correcting some errors, the most accurate means of determining gene transcript behavior will require empirical evidence. When DNA microarrays are designed to assist the statistical analysis, best practice can be achieved. For example, the replicate printing of a core control group of elements in several locations throughout an array would greatly simplify detection of spatial bias. Also, doping controls that consist of a class of non-homologous RNA transcripts could serve as independent verification of normalization methods. Finally, the correlation of expression amongst elements with high degrees of cross-homology does not prove cross-hybridization. In summary, the combination of careful array design, empirical verification and accurate mathematical models of error will result in the best practice of microarray data analysis.

ACKNOWLEDGEMENTS

We would like to acknowledge the National Science Foundation (NSF) for providing functional genomics funding (grant number 9872638) and through its postdoctoral fellowship program in bioinformatics (David Finkelstein).

REFERENCES

Kruskal WH, Wallis WA: The use of ranks in one-criterion variance analysis. Journal of the American Statistical Association: 47: 583-621 (1952).

Neter J, Wasserman W, Kutner MH: Applied Linear Statistical Models. Irwin, Homewood, Ill (1990).

Spellman PT, Sherlock G, Zhang MQ, Iyer VR, Anders K, Eisen MB, Brown PO, Botstein D, Futcher B: Comprehensive identification of cell cycle-regulated genes of the yeast Saccharomyces cerevisiae by microarray hybridization. Mol Biol Cell 9: 3273-97. (1998).

Tusher VG, Tibshirani R, Chu G: Significance analysis of microarrays applied to the ionizing radiation response. Proc Natl Acad Sci U S A 98: 5116-21. (2001).

Yang YH, Dudoit S, Luu P, Speed TP: Normalization for cDNA Microarray Data UC Berkeley Tech Report (2000).

5

A METHOD TO IMPROVE DETECTION OF DISEASE USING SELECTIVELY EXPRESSED GENES IN MICROARRAY DATA

Virginie Aris and Michael Recce
Center for Applied Genomics, Public Health Research Institute, and Center for Computational Biology and Bioengineering, NJIT

Abstract: We describe a method to improve the classification of microarray data presented in Golub et al. (1999) through the analysis of present versus absent calls in selectively expressed genes. This method does not rely on scaling or normalization factors in the comparison of data across subjects. Several genes in the Golub et al. (1999) dataset are selectively expressed between acute myeloid leukemia (AML) and acute lymphoblastic leukemia (ALL). We show that the presence or absence of expression in the 30 to 100 most selective genes is sufficient to correctly classify and diagnose the disease state of the subjects in the training dataset, and to with only one error in the independent set. In this initial analysis, the level of gene expression is not used. The exemplar, or cluster center, for each of the two diseases is computed as the real-valued average of the (expressed/not expressed) binary values for each of the most selective genes of the subjects with each disease (27 ALL, 11 AML). The Euclidian distance to each of the two exemplars is then computed for each subject. Members of a cluster are closer (smaller in distance) to the exemplar of that cluster than to the exemplar of another. For example, the range of distances of the 10 most selective genes from the AML subjects in the training set to the AML exemplar is 0.05 to 0.28, and to the ALL exemplar is 0.62 to 0.87. These data, along with the distances of the ALL subjects from the two exemplars show that the two clusters are well separated, with no overlap.

Key words: selective expression, normalization, shuffling

INTRODUCTION

Microarray technology is likely to have a substantial impact on cancer research, both as a means for elucidating biochemical pathways, as a guide for drug design and discovery, and as a method for diagnosis of the stage and state of disease. However, application of this technology depends on accurate comparison of the level of gene expression across a set of microarrays. In general this process requires normalization or scaling of measurements made on each microarray, so that differential changes in gene expression can be observed. In the most frequently used normalization method, the data are scaled by the ratio of the mean of the frequency distribution of gene expression levels on one microarray to that on a control microarray. This method works well as long as the gene expression distributions have a similar form on all microarrays, but there is little evidence to support the validity of this assumption. Furthermore the dependence on the accuracy of this normalization process is exacerbated by the magnitude of the corrections, which often increase or decrease measured expression level values by over one hundred percent.

This paper presents an alternative method for comparison of gene expression across a set of microarrays that is not dependent on the standard normalization process. Instead of using differential gene expression, we evaluate the usefulness of selective expression. Selectively expressed genes are generally present in microarrays from one group, and are less likely to be present in the group that is being distinguished from it. This method was evaluated on data from the Golub et al. (1999) study, whose aim was to classify and predict classes of leukemia by determining which genes have expression levels that are most correlated with different disease states. Their analysis method was based on the differential expression levels of all the genes on the microarrays. The expression data was derived from bone marrow and peripheral blood samples from patients suffering with acute lymphoblastic leukemia (ALL) and acute myeloid leukemia (AML). The acute lymphoblastic leukemia comprised of two subtypes: T-cell and B-cell leukemia. The dataset was split into two sets, a training set and an independent evaluation set. The training set was used to develop a statistical separation of the two groups, and was composed only of bone marrow samples. The independent set was then used to evaluate the separation, but some of the independent set samples were derived from peripheral blood and therefore added heterogeneity to the analysis.

All of the samples were hybridized on Genechips, produced by Affymetrix Inc., one the most frequently used types of microarrays. Genechips are high-density oligonucleotide probe arrays made with photolithographic techniques. Each probe cell contains 25mer oligonucleotides complementary to a specific gene sequence in the case of a perfect match probe cell (PM), or complementary except for a homomeric base mismatch at the central position for the mismatch probe cell (MM) [Lockhart et al., 1996]. The Affymetrix software employs a decision matrix based on metrics comparing the intensity of the PM to the MM to determine if a transcript is present (P), marginal (M) or absent (A). The Golub et al. study did not take into consideration the present and absent calls. Using a neighborhood analysis Golub at al. (1999) were able to classify 36 of the 38 samples in the training set and 29 of the 34 independent samples. They also used self-organizing maps [Kohonen, 1990] for automatic discovery of the classes. In contrast, we used the present and absent calls as an indicator of selective expression, and evaluated the extent to which these calls provide sufficient information to separate the two types of leukemia.

METHODS

In the initial analysis, the level of gene expression is not used. The samples in the training set are placed into two groups (27 ALL, 11 AML) and the absent and present calls of each of the genes in each group are converted into binary numbers, with one corresponding to present and zero corresponding to absent. We also considered the marginal genes and the present genes with an average difference below zero to be absent. The selectivity of each of the genes is computed as the absolute value of the difference between the real-valued average of the (expressed/not expressed) binary values for each of the two disease groups. With this metric, a gene that is present in all of the samples in one group and absent in all of the samples in the other group is maximally selective, with a selectivity of one. In contrast a gene that is absent in both groups, present in both groups, or present in the same percentage in each of the two groups is not selective at all, and has a selectivity of zero. A gene would be considered significant if it is twice as likely to be expressed in one group as in the other. In other words a gene is considered significant if it has a selectivity, or absolute difference, that is larger than 0.5. The group average represents the ideal behavior of a sample from this group and is called an exemplar. The 7129 genes are sorted into a ranked list from the most selective to the least selective, and the most selective genes are used to construct an exemplar vector for each of the two groups. The dimensionality of the exemplar vector is set by the number of genes that have been included in its definition.

The likelihood of selective genes occurring by chance was evaluated by randomly shuffling the data. The samples were randomly assigned to the two groups, which had the same size as the true AML/ALL groups and the same process was used to construct a ranked list of selective genes. The average selectivity for each rank on this list was computed for 45 shuffles of the data, and this is compared with the selectivity of the genes in the true AML/ALL groups.

The usefulness of selective genes as a method for classifying samples, was evaluated by computing the Euclidian distance from each microarray to each of the two exemplar vectors. These distances were computed for all of the microarrays in both the training and independent sets. Members of a group should be closer (smaller distance) to the exemplar of that group than to the exemplar of the other group. The exemplars computed from the training set are used to classify the data from the independent set.

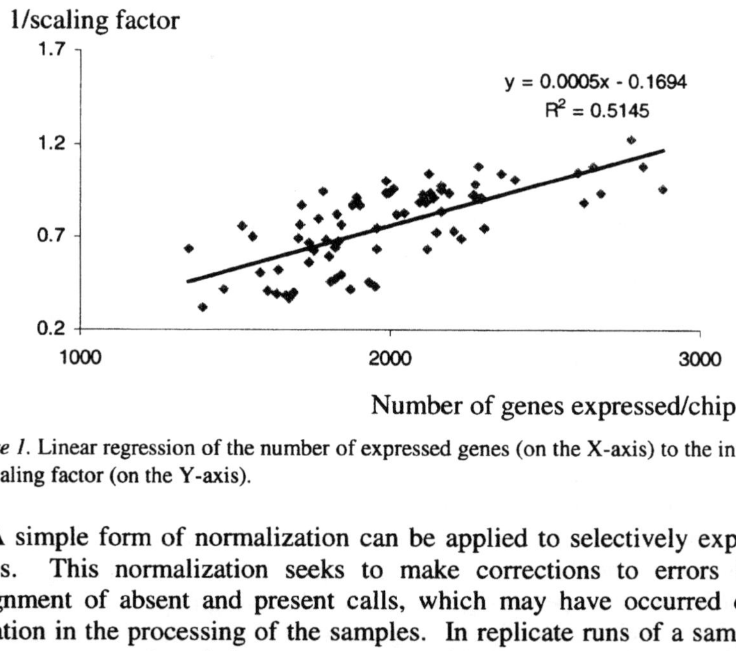

Figure 1. Linear regression of the number of expressed genes (on the X-axis) to the inverse of the scaling factor (on the Y-axis).

A simple form of normalization can be applied to selectively expressed genes. This normalization seeks to make corrections to errors in the assignment of absent and present calls, which may have occurred due to variation in the processing of the samples. In replicate runs of a sample on microarrays, we found that some genes with low expression levels have absent calls on chips with low hybridization or high background (data not shown). Those samples that are more successfully processed might be more likely to have more present calls. This hypothesis is supported by the correlation shown in Figure 1. This figure shows the relationship between the number of genes with present calls on a microarray and the scaling factor computed by the Affymetrix software. There is a linear relationship between the inverse of the scaling factor and the number of genes expressed in the chips (R^2= 0.51). Therefore chips with fewer genes expressed require a larger scaling factor, and chips with more genes expressed have a smaller scaling factor. In general, higher scaling factors might be needed if the processing of the microarray was less successful. This implies that microarrays with lower average expression levels might have genes that should be present, but their expression level is not discernable from the background. In some sense the Affymetrix normalization scales low expression levels, but it is impossible to scale up an absent call. Instead we scale down the microarrays with more genes expressed by setting the genes with lowest expression level to absent. This method is a statement that we trust the comparisons of expression levels within a microarray more than we trust the comparisons of expression levels across microarrays.

More precisely, the expression levels of all of the genes in each microarray are ranked from the highest expression level to the lowest expression level. The gene with the lowest ranked expression level on the microarray with the largest number of genes is set to absent and the selectivity of the genes is computed. This process can be continued until the selectivity starts to decrease or until all of the chips have at least as many genes present as the average number of genes present at the start of the process. The selectivity of a gene increases if it is usually absent in one group, usually present in the other group, and one or more of the present calls in the usually absent case have low expression values on microarrays with a larger than average number of genes expressed. For example, a gene that is absent in all of the ALL microarrays except for one, and is present in the all of the AML microarrays, has a selectivity that is less than one. If the one ALL present call has a low expression level on a microarray with more than the average number of present genes, then this normalization will set the ALL present call to absent, and the selectivity will increase to be exactly one.

In the initial form of the training data the number of present calls ranges from 1352 to 2877. The normalization is applied in the first instance keeping the highest ranked 2500 expression levels from each of the microarrays. Any of the microarrays that has more than 2500 genes present has the genes with the lowest expression levels now forced to be absent. The genes with selectivity greater than 0.5 prior to this process are called the selective genes. We compute the difference between the number of selective genes that increase their selectivity and the number of selective genes that decrease their selectivity. The same process is repeated in which the highest ranked 2250, 2000, 1750, 1500, and 1352 genes are kept as present, and find the number of genes that optimizes the selectivity of the selective genes.

RESULTS

Figure 2 shows the distribution of absent and present calls in this dataset. The fact that a substantial fraction of the genes are absent suggests that selective expression might usefully distinguish between groups of samples represented in the data. If the majority of the genes are always present then the data must be normalized to look for differential expression. Also, there is a large variation in the number of genes present from one slide to another, from 1352 genes found present in sample 27 to 2890 in sample 5, with an average of approximately 2000 out of over 7000 genes.

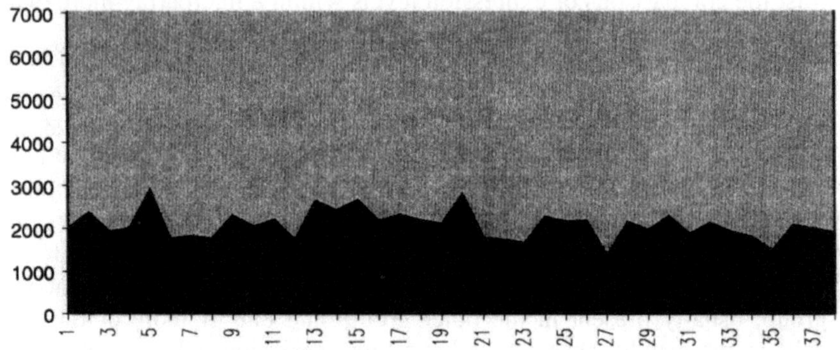

Figure 2. Number of genes Present ■ or Absent ■ across the 38 training samples (on the X-axis).

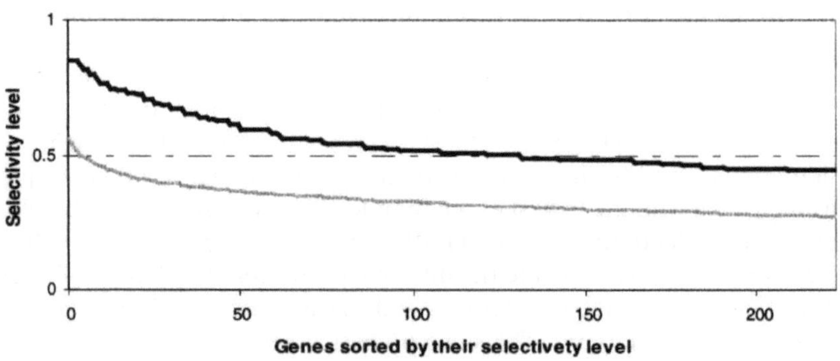

Figure 3. The selectivity level is the absolute difference between the ALL exemplar and the AML exemplar ■ The selectivity level between the shuffled samples ■ is represented as a comparison.

When the selectivity metric described above, is applied to the two leukemia groups, 121 genes are found to be selective. This included several genes that were also found to be important in the distinction between the two types of Leukemia by Golub et al. (1999), including MB-1, zyxin, HOXA9, and cyclin D3 (Table 1). In the list of the 24 most selective genes we see that some genes, such as the myosin light chain, are more expressed in the ALL leukaemia. Others, such as zyxin, are expressed more often in the AML group. As described in the Methods section, this selectivity was compared to 45 random shuffling of the samples into the two groups. Those selective genes had significantly higher difference in expression than the randomly shuffled sets (last column on the table). Figure 3 shows the selectivity of the genes, after ranking. The x-axis is the position of the gene in this ranked list, and the y-axis is the selectivity.

Table 1. The 24 genes that are the most selectively expressed from the AML/ALL training set, with their average expression for the ALL/AML groups, the absolute difference, compared to the absolute difference of shuffled set and number of standard deviation (No. SD) of the AML/ALL selective genes away from the average shuffled groups.

	ALL-AML grouped selectively				**Shuffle comparison**		
Rank	Gene Description	Average ALL	Average AML	Absolute Difference	Mean Absolute Difference	Standard Deviation (SD)	No. SD
1	DAGK1 Diacylglycerol kinase	0.85	0	0.85	0.60	0.07	3.51
2	CYSTATIN A	0.15	1	0.85	0.56	0.05	5.36
3	KIAA0035 gene	0.85	0	0.85	0.54	0.05	6.25
4	MYL1 Myosin light chain	0.93	0.09	0.84	0.52	0.05	6.61
5	LEPR Leptin receptor	0.19	1	0.81	0.51	0.04	6.98
6	NUCLEOLYSIN TIA-1	0.81	0	0.81	0.50	0.04	7.83
7	Terminal transferase mRNA	0.89	0.09	0.8	0.49	0.04	7.59
8	Inducible protein mRNA	0.89	0.09	0.8	0.48	0.04	8.42
9	CHRNA7 Cholinergic receptor	0.04	0.82	0.78	0.48	0.04	8.33
10	PPBP	0.15	0.91	0.76	0.47	0.04	8.04
11	CST3 Cystatin C	0.15	0.91	0.76	0.47	0.04	8.01
12	RB1	0.85	0.09	0.76	0.46	0.03	8.51
13	Zyxin	0.07	0.82	0.74	0.46	0.03	8.19
14	BTF5	0.93	0.18	0.74	0.46	0.03	8.33
15	GLUTATHIONE S-TRANSFERASE	0.26	1	0.74	0.45	0.03	8.58
16	MB-1	0.74	0	0.74	0.45	0.03	8.77
17	KIAA0230	0.74	0	0.74	0.44	0.03	9.06
18	adipsin	0	0.73	0.73	0.44	0.03	8.81
19	PTX3	0	0.73	0.73	0.44	0.03	8.97
20	IGIF	0	0.73	0.73	0.43	0.03	9.26
21	RABAPTIN-5	0.81	0.09	0.72	0.43	0.03	9.25
22	UBIQUITIN-LIKE PROTEIN GDX	0.81	0.09	0.72	0.43	0.03	9.15
23	HOXA9	0.11	0.82	0.71	0.43	0.03	8.61
24	Cyclin D3	0.89	0.18	0.71	0.42	0.03	8.58

To be able to classify the samples in one group or the other, we computed the Euclidian distance of their binary call (1 for present and 0 for absent) to the exemplar. We then averaged this distance among the 10, 20, 30 most selective genes and determined to which exemplar that sample was the closest. With the ten most selective genes we were able to completely separate the 2 groups in the training set (Figure 4).

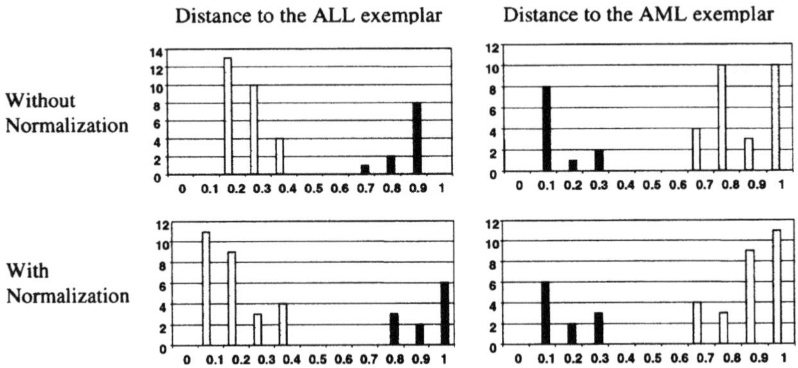

X-axis: Distance to the exemplar, ALL on the left and AML on the right.
Y-axis: Frequency distribution

Figure 4
Frequency distribution of the distance of the independent samples to the AML and ALL exemplar before and after normalization using the 10 most selective genes on the training set. The histogram in each panel has one entry for each sample, and a different color for the two types of leukemia ALL samples and AML samples ■.

After applying the normalization process described above, we found that the selectivity was maximized by retaining, as present, the 2000 genes on each microarray that have the highest expression level on that microarray. The lower two panels of Figure 4 show that this normalization improves the separation of the two groups. Using the ten most selective genes, the AML samples are closer to the AML exemplar and that the ALL samples are further from the AML exemplar.

Figure 5 shows the effect of changing the dimensionality of the exemplar vector on cluster separation. The prior analysis was based on the ten most selective genes, or equivalently on ten dimensional exemplar vectors. The x-axis of this figure is the dimensionality of the exemplar vector. Again the included genes have been ranked, and the genes are added to the exemplar in rank order from the most selective to the least selective. The y-axis is the distance of each of the samples from the ALL exemplar. Each sample is plotted as a distinct line. The top panel shows the distances prior to applying the normalization process and the lower panel shows the distances after applying the normalization process. Over most of the range shown in the figure the normalization process has increased the separation of the two groups. As more genes are included in the exemplar vector the graphs gradually move closer together.

A

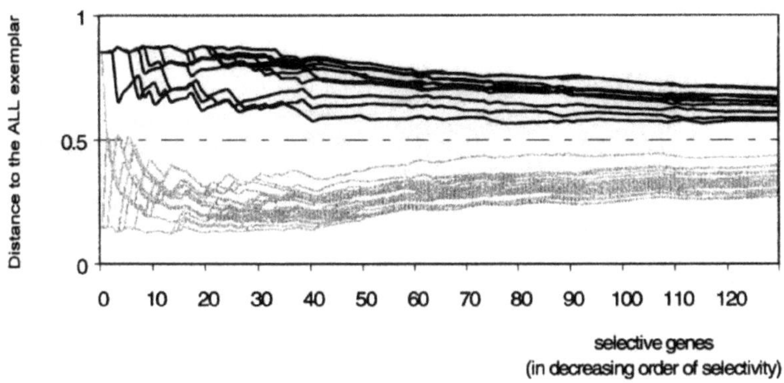

B

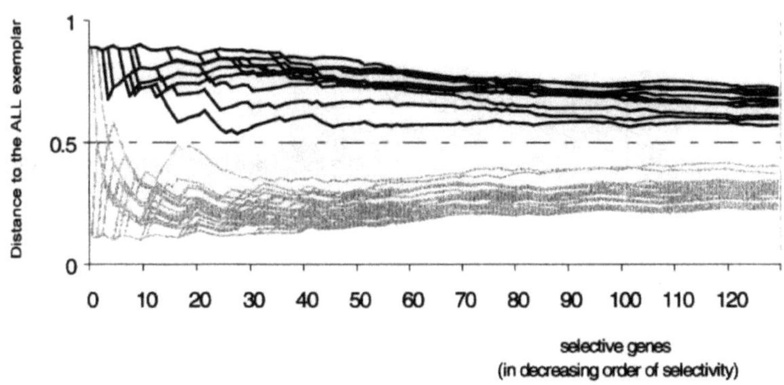

Figure 5. ALL samples ▒ and AML samples ■ (A) Distance of the training set samples to the ALL exemplar before normalization. (B) Distance of the training set samples to the ALL exemplar after normalization.

The same analysis was applied to the independent set, but the distances were computed to the exemplar that was computed from the training data. As shown in Figure 6, this technique was successful at separating the two groups. Only one sample, number 66, was incorrectly classified. The top panel shows the distances of the two groups from the ALL exemplar prior to normalization, and the lower panel shows the distances after the same normalization as had been applied to the training set.

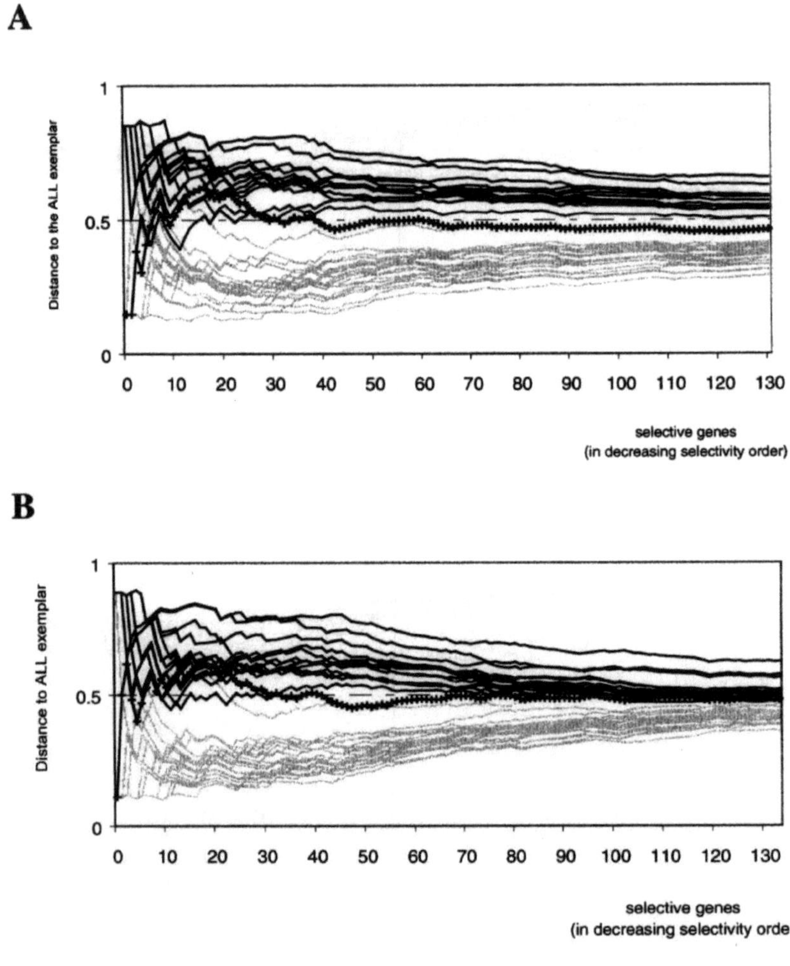

Figure 6. ALL samples ▒ and AML samples ■, +++ misclassified sample number 66, (A) Distance of the independent set samples to the ALL exemplar before ranking. (B) Distance of the independent set samples to the ALL exemplar after ranking.

In this normalization no additional optimization was performed, and as before the 2000 genes with highest expression level from each microarray were kept as present. Note that again the normalization increases the separation of the two groups.

The same method was used to examine other possible classifications of the data. There was information in the data about the sub grouping of the ALL samples into T-cell and B-cell type leukemia. Just as was found in the ALL-AML classification, genes were identified in the T-cell to B-cell groups that are more selective than expected by chance, as evaluated by shuffling the samples. The resulting exemplar vectors were able to completely separate the T-cell and B-cell groups in the training set without error. In the independent set, one out of the 20 ALL independent samples, number 67, which was a peripheral blood sample, was misclassified. The method was also applied to the classification of success and failure of treatment and to female and male subjects. In both of these two cases there was no significant difference between the selectivity of genes found in the real groups from that found in the randomized groups. The absence of a correlation in the treatment response case may be due in part to the small number of samples (11 in the training set).

CONCLUSION

Selective gene expression is sufficient for separating the AML vs. ALL samples, and T-cell vs. B-cell samples. Selective expression does not have the same dependency as differential expression on normalization of microarray data, and for this reason it may be more robust regardless of variations in processing. Some of the most selective genes were differentially expressed in the Golub et al. (1999) study (e.g. MB-1, zyxin, HOXA9, cyclin D3). Other genes like Catalase (X04085) which were differentially expressed in both ALL and AML were not selected by this method, and genes such as TCL1 gene (T cell leukemia) were selective but were not found to be differentially expressed in the prior study.

The shuffling method provided a useful verification of the use of selective genes to classify the data into groups. The AML and ALL groups, and the T-cell and B-cell groups were far from the randomly constructed datasets, while the success–failure classification and female–male classification was close to chance.

Our approach is orthogonal to prior studies and complementary, since it focuses on genes that are selectively expressed rather than on differential expression levels. This simple approach is very robust, powerful and easy to use in diagnostic microarray development as it contains strongly expressed genes. A synthesis of the qualitative and quantitative methods should improve the classification performance, and add understanding to the mechanism of action of the genes involved in some pathology. The software will be publicly available for academic institutions at this site: http://www.cag.icph.org.

REFERENCES

Golub TR, Slonim DK, Tamayo P, Huard C, Gaasenbeek M, Mesirov JP, Coller H, Loh ML, Downing JR, Caligiuri MA, Bloomfield CD, Lander ES. Molecular classification of cancer: class discovery and class prediction by gene expression monitoring. Science 286(5439): 531-537, 1999.

Kohonen, T., 1990. The Self-Organizing Map. Proceedings of the IEEE, 78: 1464-1477.

Lockhart D. J., H. Dong, M C. Byrne, M. T. Follettie, M. V. Gallo, M. S. Chee, M. Mittmann, C. Wang, M. Kobayashi, H. Horton, and E. L. Brown, 1996. Expression monitoring by hybridization to high-density oligonucleotide arrays. Nat. Biotechnol., 14:1675-1680

6

COMPUTATIONAL ANALYSIS OF LEUKEMIA MICROARRAY EXPRESSION DATA USING THE GA/KNN METHOD

Leping Li,[1] Lee. G. Pedersen,[2] Thomas A. Darden[3] and Clarice R. Weinberg[1]

[1] Biostatistics Branch and [3]Laboratory of Structural Biology, National Institute of Environmental Health Sciences, Research Triangle Park, North Carolina 27709, USA
[2]Department of Chemistry, University of North Carolina, Chapel Hill, North Carolina 27709, USA

Abstract: We recently developed a multivariate method that selects a subset of discriminative genes for sample classification based on gene expression data. The method combines a search tool, a genetic algorithm (GA), and a non-parametric pattern recognition method, based on the k-nearest nearest neighbors (KNN) approach. We begin by selecting many subsets of genes that can discriminate among classes of samples using a training set. Subsequently, the genes are ranked according to the frequency of gene selection. The top-ranked genes (e.g. 50) are then used to classify test set samples. For a widely-available set of leukemia data, the top 50 genes identified by the GA/KNN method not only correctly classified 33 of the 34 test set samples, but also discovered the two distinct clinical subtypes within ALL without applying prior knowledge. The method has been successfully applied to several expression data sets. It may be used to identify a subset of informative genes (biomarkers) for sample classification for a variety of profiling studies including tumors.

Key words: leukemia, pattern recognition, genetic algorithm, k-nearest neighbors, variable selection, principal component analysis, microarray

INTRODUCTION

Microarray technology has made it possible to monitor the global gene expression patterns of cells. It has been used to profile the gene expression of normal and transformed human cells in several tumors including colon

[Alon et al., 1999], leukemia [Golub et al.1999], breast [Perou et al., 2000], lymphoma [Alizadeh et al., 2000], prostate [Bull et al., 2001], and melanoma [Bittner et al., 2000] as well as the NCI's 60 human tumor cell lines [Ross et al., 2000]. These studies may shed light on the mechanisms of cell maltransformation. Therefore, they may be useful in elucidating the mechanisms involved in carcinogenesis and in identifying characteristic patterns for cancer diagnosis and classification.

Methods that mine large expression data sets for sample classification have been reported [Golub et al., 1999; Ben-Dor et al., 2000; Bittner et al., 2000; Perou et al., 2000; Li et al., 2001a & 2001b]. The GA/KNN method that we developed is a multi-dimensional classification method. It not only selects a small subset of genes that *jointly* discriminates among classes of samples (e.g. normal vs. tumor) but also assesses the relative predictive importance of all genes for sample classification [Li et al., 2001a & 2001b]. We begin by selecting a subset of genes that can discriminate between different classes of samples. When many such subsets of differentially expressed genes are obtained independently using the GA/KNN method, the relative importance of genes for sample classification can be assessed by examining the frequency of selection of the genes into these near-optimal subsets. The most informative set of genes (class predictor) is subsequently used to classify an "unknown" (validation) set of samples. We have applied the method to several data sets including a colon data set [Alon et al., 1999], a lymphoma data [Alizadeh et al., 2000], and a toxicogenomics data set generated at the National Institute of Environmental Health Sciences (to be published). Herein we report the detailed analysis of the leukemia data [Golub et al., 1999] using the GA/KNN method to find sets of genes that can distinguish between acute lymphoblastic leukemia (ALL) and acute myeloid leukemia (AML).

METHODS

Data set

The original leukemia data were downloaded from the web at http://www.genome.wi.mit.edu/MPR. The data contain the expression levels of 6,817 genes across 72 samples, of which 47 had been classified as ALL and 25 as AML [Golub et al., 1999]. We excluded any genes with expression levels below 50 in more than 80% of the 72 samples. The number of genes remaining after this filtration is 5,545. The expression values were then transformed as follows:

$$f(x) = \begin{cases} c + \log_{10}(x) & x > 1 \\ c & x \geq -1 \text{ and } x \leq 1 \\ -c - \log_{10}(\text{abs}(x)) & x < -1 \end{cases}$$

where x is the expression level, c is a constant and was set to 0.

We divided the data set into a training set (first 38 samples) and a test set (34 samples) following Golub et al. [1999]. The training set was used to obtain a subset of genes that can discriminate between AML and ALL. The 50 most informative genes obtained using the training set were subsequently used in validation, to predict the classifications of the test set samples.

The GA/KNN method

Details of the GA/KNN method have been reported [Li et al., 2001a and 2001b] and are available on the web at the following location: http://dir.niehs.nih.gov/microarray/datamining. The method employs a non-parametric pattern recognition method, the k-nearest neighbor (KNN) [see e.g., Vandeginste et al., 1998], and a searching tool, which is a genetic algorithm (GA) [for a review, see Judson 1997]. The KNN method was used as the classification tool that distinguishes between discriminative and non-discriminative sets of genes while the GA was used to choose a relatively few subsets of genes (from many possible combinations) for testing.

In brief, each sample is represented by a pattern of expression that consists of d genes. Each sample is then KNN classified according to the class memberships of its k (arbitrarily set to 3) nearest neighbors, as determined by the Euclidean distance in the d-dimensional space. If all of the 3 nearest neighbors of a sample are ALL, the sample is classified as ALL; similarly for AML. If the 3 nearest neighbors are not of the same class, the sample is considered unclassifiable.

The GA is used to search high-dimensional space, since selecting an optimal subset of optimal genes from a large gene pool is a combinatorial problem. Initially, a population of randomly selected "chromosomes" is generated. Each "chromosome" consists of d genes. The "fitness" (merit) of each "chromosome" is determined by its ability to classify the training set samples. If the class memberships of a training set sample and its three nearest neighbors in the particular d-dimensional space defined by a "chromosome" agree, a score of 1 is assigned to that sample. These agreement scores are summed across the training set, and we refer to this sum, divided by the number of samples (here 38), as the fitness score. Thus, the GA is used to optimize the fitness score.

A set of *d* genes was considered discriminative in which each sample and its three nearest neighbors are of the same class for at least 37 of the 38 training set samples. To study the sensitivity of gene selection to the choices of *d*, we systematically examined various sizes of *d* (e.g., 5, 10, 20, 30, 40, 50, 70, and 100). For each *d*, 20,000 discriminative sets of genes (referred to as the near-optimal "chromosomes") that potentially discriminate between AML and ALL were obtained. Subsequently, the genes were ranked according to the frequency of selection into near-optimal "chromosomes".

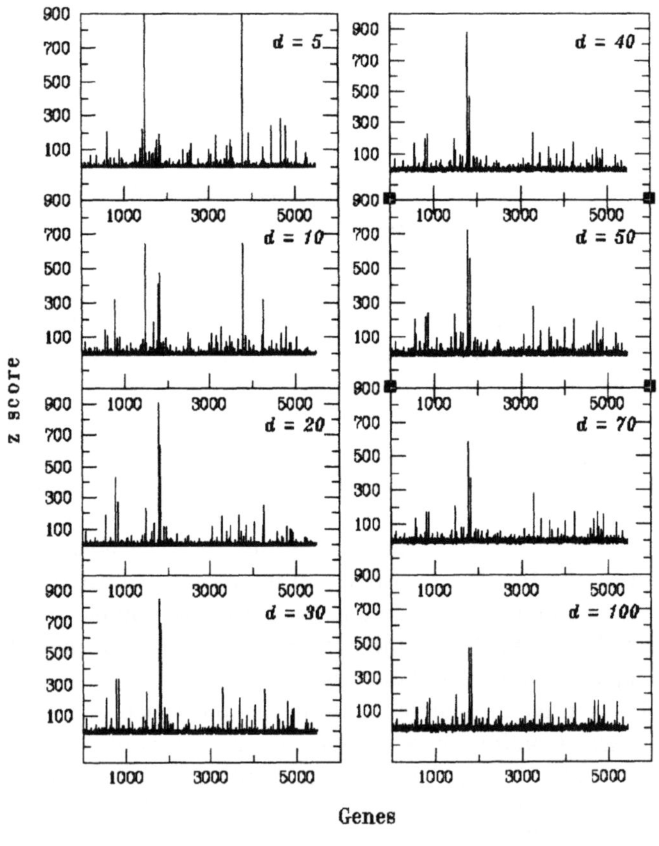

Figure 1. The statistical z-score with which each of the 5,455 genes is selected among the 40,000 solutions. Let $Z = [S_i - E(S_i)]/\sigma$, where S_i is the number of times $gene_i$ is selected, $E(S_i)$, is the expected number of times $gene_i$ is selected, σ is the square root of the variance. Let, $A = 40{,}000$, $P(gene_i) = d/5455$, the probability of $gene_i$ being selected (if random). Then, $E(S_i) = P(gene_i) \cdot A$, and $\sigma = \sqrt{\{P(gene_i) \cdot [1 - P(gene_i)] \cdot A\}}$. The two highest peaks in $d = 5$ were truncated.

The most frequently selected genes (e.g. top 50) were then used to classify the test set samples.

RESULTS

Frequency of gene selection and choice of *d*

The statistical *z* score, based on normalizing the frequency with which each of the 5,455 genes was selected in the 20,000 solutions for each *d*, is shown in Figure 1. A few genes dominate the selection when *d* is small (e.g. 5). As *d* increases, the pattern of gene selection stabilizes. The correlation between the *z* scores for two choices of *d* is shown in Figure 2.

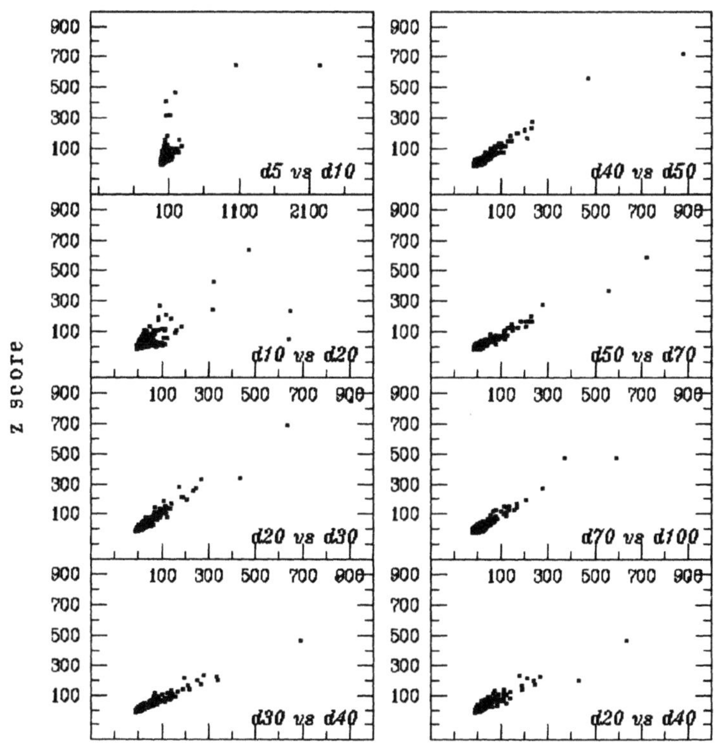

Figure 2. Correlation between the *z* scores for two consecutive choices of *d*. The *z* scores were calculated as described in Fig. 1.

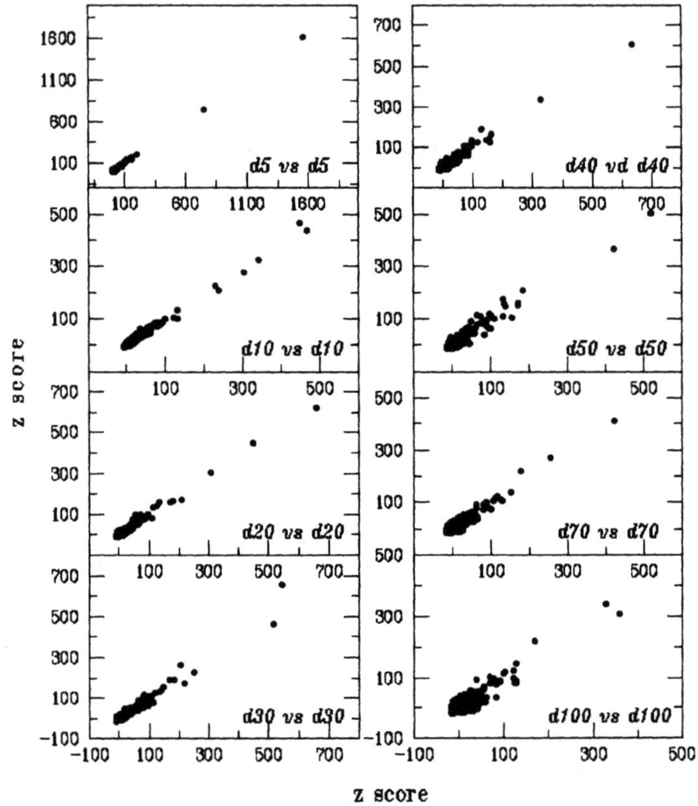

Figure 3. Reproducibility of gene selection as represented by the correlation between the *z* scores from two independent runs for the same *d*. For each run, 20,000 solutions were obtained for each *d*.

To examine the reproducibility of gene selection, an additional 20,000 near-optimal "chromosomes" were obtained independently for each *d*. Again, each near-optimal "chromosome" contains *d* genes that can discriminate between AML and ALL for the training set samples. The correlation between the *z* scores for two independent runs for each *d* is shown in Figure 3. The reproducibility is apparently high for all d except *d* = 100. For large *d* (e.g. 100), gene selection is likely contaminated with noise. Furthermore, the calculation becomes computationally expensive as *d* increases. Taken together, a choice of *d* between 20 and 70 appears to yield a stable gene selection for the leukemia data set. We chose a *d* of 40, taking reassurance from the fact that the selection of optimal genes is evidently insensitive to this choice.

Top genes

The genes were ranked based on the frequency of gene selection (d = 40) and the top 50 genes are listed in Table 1. It is noteworthy that many more were selected with high z scores. The complete list of 5,455 genes based on frequency rank is available on the web at this address: http://dir.niehs.nig.gov/microarray/datamining/.

Among the top 50 genes, several were myeloid, T-lymphocytic- or B-lymphocyte-specific antigens or surface markers. For instance, CCAAT/enhancer binding protein δ (C/EBPδ) is a transcriptional regulator involved in myeloid cell differentiation [Sterneck et al., 1998]. Myeloperoxidase (MPO) is a well-established marker of myeloid differentiation [Austin et al., 1998; Winterbourn 2000]. The CD2 is a T-cell antigen that has been used as a diagnosis marker for T-cell ALL [Manabe et al., 1998]. It plays an important role in mediating the interactions between human T lymphocytes and accessory cells [Brown et al., 1989]. The B-cell lymphocyte kinase, Blk, is a src-family protein tyrosine kinase that is expressed preferentially in B lineage cells [Malek et al., 1998]. Blk may play an important role in B cell proliferation [Malek et al., 1998].

Immunoglobulin-associated beta (B29), also called CD79b, belongs to a family of surface adhesion molecules on B lymphocytes. B29 and the associated transmembrane protein mb1 are crucial for assembly and membrane display of the B-cell receptor [Hombach et al., 1990]. Mutation of *B29* has been associated with chronic lymphocytic leukemia [Gordon et al., 2000]. The octamer binding factor 1 (OBF-1) is a B-lymphocyte-specific transcription factor [Schubart et al., 1996; Matthias, 1998]. It has been suggested that over-expression of *OBF*-1 might play a role in the pathogenesis of germinal center-derived B cell lymphoma [Greiner et al., 2000]. CD19 is also B lymphoid antigen and used as a diagnosis marker for

B-cell ALL [Scheuermann & Racila, 1995]. The *TCL1* is an oncogene located on chromosome 14 band q32.1 that has been implicated in the development of mature T cell leukemia [Virgilio et al., 1993; Pekarsky et al., 2000].

Several oncogenes or tumor antigens were also frequently selected, for instance, the proviral integration oncogene *spi1*, *GRO2* oncogene, and carcinoembryonic antigen precursor gene.

Table 1. The 50 most frequently selected genes[a]

Student's *t*-statistic[b]	*z*-score	Gene bank Accession No.	Gene name
7.470	1010.517	M84526	DF D component of complement (adipsin)
-3.892	718.387	M89957	IGB Immunoglobulin-associated beta (B29)
4.152	323.952	X52056	SPI1 oncogene spi1
3.973	323.713	HG4321-HT4591	Ahnak-Related Sequence
-4.235	308.496	J03589	UBIQUITIN-LIKE PROTEIN GDX
4.666	285.432	M27891	CST3 Cystatin C
-3.464	273.374	X58072	GATA3 GATA-binding protein 3
-2.579	263.804	M16336	CD2 CD2 antigen (p50),
-5.516	257.200	D87076	KIAA0239 gene, partial cd
-5.675	243.754	M86406	ACTN2 Actinin alpha 2
-3.906	232.557	X82240	TCL1
-3.860	189.013	Z49194	OBF-1
5.791	177.720	M83667	NF-IL6-beta protein mRNA
-4.530	167.719	X66533	Guanylate cyclase, β1
-3.069	153.699	M63488	RPA1 Replication protein A1 (70kD)
3.155	145.181	M63438	GLUL Glutamate-ammonia ligase
3.925	141.832	M59820	CSF3R Colony stimulating factor 3 receptor
-3.796	140.492	M29540	Carcinoembryonic antigen precursor
-3.728	136.568	U90902	Clone 23612 mRNA sequence
2.212	129.917	S34389	HMOX2
-2.627	128.673	X99920	S100 calcium-binding protein A13
-3.416	124.270	U89922	LTB Lymphotoxin-beta
-3.246	117.284	S76617	BLK Protein-tyrosine kinase blk
-5.048	117.284	U16954	(AF1q) mRNA
-5.265	117.093	X52142	CTPS CTP synthetase
4.197	115.227	M19507	MPO Myeloperoxidase
4.170	115.083	HG3921-HT4191	Homeotic Protein C6, Class I
3.783	113.265	D87433	KIAA0246 gene, partial cds
2.980	107.331	M26679	HOMEOBOX PROTEIN HOX-A5
-5.190	98.575	AF009426	Clone 22 mRNA
-3.346	98.383	L16991	DTYMK Deoxythymidylate kinase
4.205	95.847	Y00339	CA2 Carbonic anhydrase II
4.686	91.493	M57731	GRO2 GRO2 oncogene
2.968	90.679	HG180-HT180	Ahnak-A Nucleoprotein Ahnak-A
-4.963	88.622	HG4755-HT5203	Spinal Muscular Atrophy 4
-3.270	87.856	X87613	Skeletal muscle abundant protein
4.529	87.282	M27783	ELA2 Elastatse 2, neutrophil
-2.685	83.645	U39226	Myosin VIIA (USH1B) mRNA
-3.629	80.822	U08377	Homolog of Drosophila splicing regulator
3.230	78.621	L31881	NFIX Nuclear factor I/X
-4.488	77.042	M28170	CD19 CD19 antigen
-3.943	74.506	U90546	Butyrophilin (BTF4) mRNA
-3.893	72.735	Z14982	MHC-encoded proteasome subunit gene LAMP7-E1
3.677	70.199	X05908	ANX1 Annexin I (lipocortin I)
-3.769	69.146	M19888	SPRR1B Small proline-rich protein 1B
-5.201	69.051	S82185	GB DEF = E. coli unknown mRNA
-3.484	67.328	D38522	KIAA0080 gene, partial cds
2.988	66.897	X61587	ARHG Ras homolog, member G (rho G)
3.751	65.366	L28821	MANA2 Alpha mannosidase II isozyme
2.986	62.591	X64594	Erythrocyte plasma membrane 50 KD glycoprotein

[a] The genes are listed in descending order based on the rank frequency obtained using the 34 training set samples and a gene-dimension of 40 (see text for details). The gene definition is adapted from (http://www.genome.wi.mit.edu/MPR). A complete list of all genes is available on http://dir.niehs.nih.gov/microarray/datamining.

[b] Student's two-sample *t*-test was performed on the 34 training set samples that were normalized to a mean and standard deviation of 0 and 1, respectively. A positive value indicates that the gene is up regulated in AML compared to ALL and conversely. The *t*-values at which P is 0.1, 0.01, and 0.001 are 1.684, 2.704, and 3.551, respectively.

Table 2. Classification of the test set samples of the leukemia data[a]

Sample	Clin.[b]	Top 25	Top 50	Top 100	Top 200	Top 500	Top 1000	Top 2000	All 5455
ALL39	0	0	0	0	0	0	0	0	0
ALL40	0	0	0	0	0	0	0	0	1
ALL42	0	0	0	0	0	0	0	0	0
ALL47	0	0	0	0	0	0	0	0	0
ALL48	0	0	0	0	0	0	0	0	0
ALL49	0	0	0	0	0	0	0	0	0
ALL41	0	0	0	0	0	0	0	0	0
ALL43	0	0	0	0	0	0	0	0	0
ALL44	0	0	0	0	0	0	0	0	0
ALL45	0	0	0	0	0	0	0	0	0
ALL46	0	0	0	0	0	0	0	0	0
ALL70	0	0	0	0	0	0	0	0	0
ALL71	0	0	0	0	0	0	0	0	0
ALL72	0	0	0	0	0	0	0	0	0
ALL68	0	0	0	0	0	0	0	0	0
ALL69	0	0	0	0	0	0	0	0	0
ALL67	0	0	0	0	0	0	0	0	0
ALL55	0	0	0	0	0	0	0	0	0
ALL56	0	0	0	0	0	0	0	0	0
ALL59	0	0	0	0	0	0	0	0	0
AML52	1	1	1	1	1	1	1	1	1
AML53	1	1	1	1	1	1	1	1	1
AML51	1	1	1	1	1	1	1	1	1
AML50	1	1	1	1	1	1	1	1	1
AML54	1	1	1	0	1	1	1	0	0
AML57	1	1	1	1	1	1	1	1	0
AML58	1	1	1	1	1	1	1	1	0
AML60	1	1	1	1	0	0	0	0	0
AML61	1	1	1	1	1	1	1	1	0
AML65	1	1	1	1	1	1	1	1	0
AML66	1	0	0	0	0	0	0	0	0
AML63	1	1	1	1	1	1	1	1	0
AML64	1	1	1	1	1	1	1	1	1
AML62	1	1	1	1	1	1	1	1	0

[a]A sample is classified as 0-ALL or 1-AML using three *training* set neighbors (2 or 3 must agree) by the KNN method using the top-ranked genes or all 5,455 genes. The genes were ranked according to the frequency of gene selection. AML66 is highlighted.
[b]Clinical classification, adapted from Golub et al. [1999].

Classification of the test set samples

The GA/KNN method correctly classified 33 of the 34 test samples using three training set neighbors (2 or 3 must agree) by KNN using the top 50 genes (Table 2). When the 100 top-ranked genes were used, a similar result was obtained except that sample AML54 was classified as ALL. As more genes were included, contaminating the system with high-dimensional noise, the number of misclassified samples increased. In fact, when all 5,455 genes were used, 9 of the 34 test set samples were incorrectly classified. This result emphasizes that not all expression data are relevant to the distinction between ALL and AML. Interestingly, nearly all the misclassified samples were AML, suggesting that the ALL samples are relatively easy to classify. Among those that were misclassified, AML66 was consistently classified as ALL (Table 2). When all 72 samples were used for training, AML66 remained classified as ALL (data not shown). Together, these results show that AML66 is an outlier.

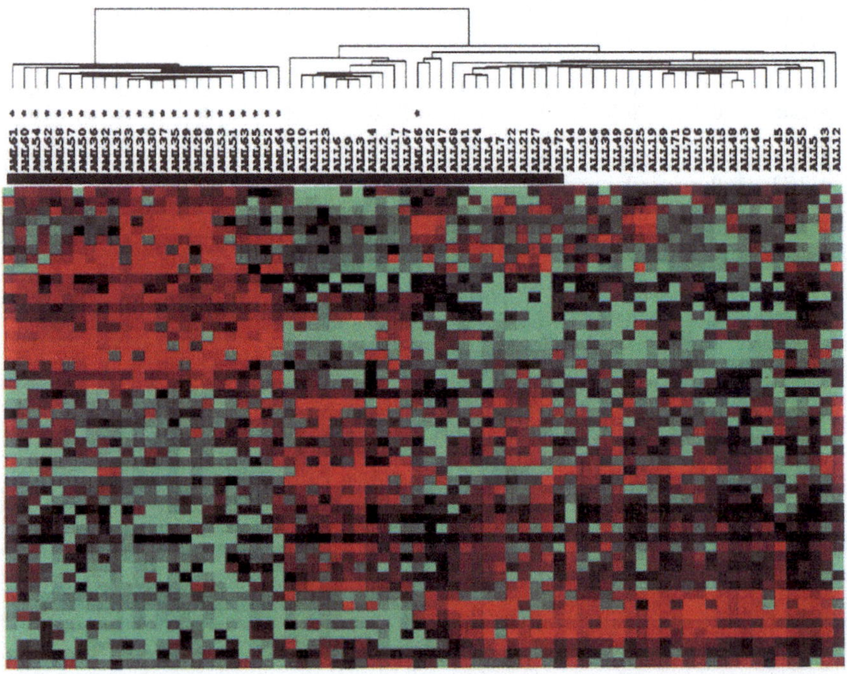

Figure 4. Clustering analysis [Eisen et al., 1998] of the 72 leukemia samples and the 50 top-ranked genes that were selected from all 5,455 genes by the GA/KNN method.

When the test set was combined with the training set, a post-processing cluster analysis of the top 50 genes using a cluster analysis program [Eisen et al., 1998] showed that the AML samples and ALL samples were clustered together correctly except AML66 (Figure 4). Furthermore, the top 50 genes found by the GA/KNN method revealed the existence of two subtypes within ALL without applying any prior knowledge. Among the 47 ALL samples, 9 have been clinically classified as T-cell ALL (ALL2, ALL3, ALL6, ALL9, ALL10, ALL11, ALL14, ALL23, and ALL67), and the remaining as B-cell ALL. When clustered using the top 50 genes, each of the 9 T-cell ALL samples were on one branch of the ALL tree together with two B-cell ALL. These results indicate that the GA/KNN method is capable of identifying genes that discriminate not only between the ALL and AML, but that the method may also unmask clinically meaningful subtypes, through subsequent cluster analysis.

We also applied principal component (PC) analysis (see e.g., Vandeginste et al., 1998] to the 72 samples using the top 50 genes. The method of single value decomposition (SVD) [Numerical Recipes, Cambridge, MA] was applied to the correlation matrix. The principal components were obtained by projecting the original data points that have been transformed (see methods) and then mean centered onto the eigenvectors. When plotting the 72 samples using the first two principal components, two distinct clusters are apparent with AML66 in the wrong cluster (Figure 5).

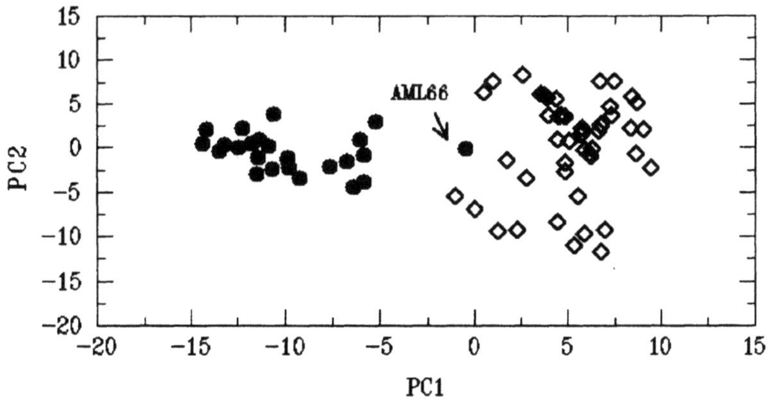

Figure 5. Principal component analysis of the 72 samples based on the 50 top-ranked genes selected by the GA/KNN method. Filled circle – AML and open diamond – ALL. Only the first two principal components are shown.

In addition, we studied whether each of the 40,000 individual near-optimal "chromosomes" could classify as well as the top 50 genes. Each test set sample was classified 40,000 times as either AML or ALL using 3 nearest *training* set neighbors by the KNN method using a majority rule (2 or 3 must agree). Interestingly, three AML samples (AML54, AML60, and AML66) were classified as ALL more than 50% of the time (Figure 6). Among the three, AML66 was misclassified 93% of the time. Furthermore, none of the 40,000 individual near-optimal "chromosomes" correctly classified more than 32 of the 34 test set samples. This result suggests that the individual near-optimal "chromosomes" are not as effective as the set of top-ranked genes (e.g. the top 50) that were selected from these "chromosomes". This instability may be due to inherent noise present in these "chromosomes".

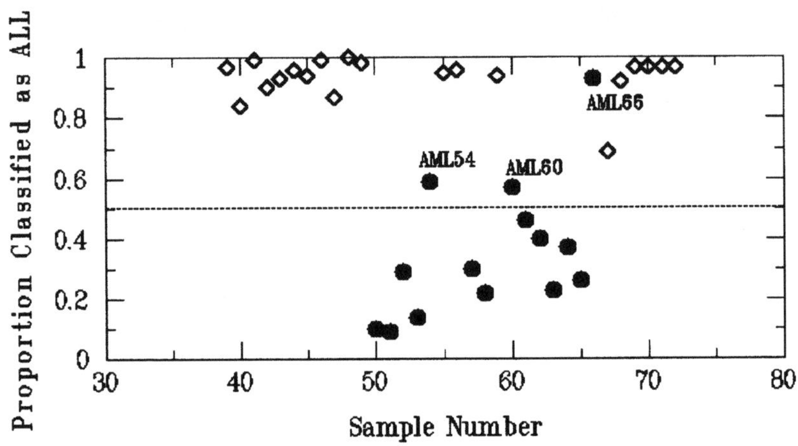

Figure 6. Percentage of being classified as ALL for the test set samples. Each test set sample was classified by three training set neighbors using the top 50 genes by KNN method. A majority rule (2 or 3 must agree) was applied. Each sample was classified 40,000 times using each of the 40,000 solutions (near-optimal "chromosomes") obtained during the training process (see text for details). Filled circle – AML and open diamond – ALL.

DISCUSSION

It is clear that not all genes are relevant to sample discrimination. Thus, identification of a subset of informative genes is important. The current approaches for selecting a subset of informative genes can be summarised by two categories, univariate and multivariate. A univariate approach examines one gene at a time whereas a multivariate approach considers several genes simultaneously. The widely used discriminative measures include the Student's *t* statistics or similar measures [Golub et al., 1999], discriminant analysis [see e.g. Vandeginste et al., 1998], neural network [Toronen et al., 1999], support vector machines [for a tutorial, see Cortes & Vapnik, 1995; Ben-Dor et al., 2000], and nearest neighbors [Li et al, 2001a & 2001b]. Alternatively, when selection of individual genes is not carried out, a few "super-genes" are constructed as the informative genes using multivariate approaches such as the principal component analysis (PCA).

Univariate approaches may fail to identify sets of genes that are discriminative *jointly*, but not singly. We have shown that the frequency of co-selection of certain sets of genes (e.g. P_{ABC} for genes A, B, and C) can be many times higher than the product of their individual frequency of selection ($P_A \times P_B \times P_C$) [Li et al., 2001a & 2001b], indicating the multivariate nature of their contribution to discrimination.

Many supervised pattern recognition methods have been employed as discrimination tools. Some methods (such as discriminant analysis) define a boundary between classes explicitly while others (such as the KNN) do so implicitly. This distinction may have an implication for sample classification, especially when samples can be heterogeneous in nature, such as tumors. It is clear that tumor subtypes may exist within known classes. Discriminative functions that can work well in the presence of such heterogeneity are preferred. Subsequently, hidden subcategories in the sample may be discovered and may prove to be etiologically or prognostically important.

The last important issue is that microarray data consist of a large number of genes (parameters) and a small number of samples and as a result, many distinct and equally effective classifiers may exist for the same training set. Most of current literature methods seek a single subset of discriminative genes. Often, the informative genes identified for a given data set vary from method to method.

The GA/KNN method selects many subsets of discriminative genes from which a single predictor set is formed by examining the frequency of gene selection. This simple approach of gene selection appears to function well. The top-ranked genes (e.g. the top 50) apparently outperform the individual subsets of genes. Furthermore, the GA/KNN method accommodates the heterogeneity within the classes, since the KNN method finds a boundary between classes *implicitly*. This unique ability of KNN facilitates subclass discovery.

In conclusion, many methods have been developed for sample classification based on gene expression data. As the quantitative aspects of the microarray technology improve and computational methods that mine the resulting large data set are developed further, the technology will have a great impact on biology, toxicology and related areas.

ACKNOWLEDGEMENTS

Part of the work was completed when LL was employed at the National Institute for Occupational Safety and Health at Morgantown, West Virginia. LL would like to thank David Umbach for advice on data transformation and insightful discussions and Shyamal Peddada for critical reading of the manuscript, and the North Carolina Supercomputing Center for access to their facilities. Most of the computations were carried out on SGI workstations.

REFERENCES

Alon,U., Barkai,N., Notterman,D.A., Gish,K., Ybarra,S., Mack,D. and Levine,A.J. *Proc. Natl. Acad. Sci. USA,*1999, *96*, 6745.

Alizadeh,A.A., Eisen,M.B., Davis,R.E., Ma,C., Lossos,I.S., Rosenwald,A., Boldrick,J.C., Sabet,H., Tran,T., Yu,X., Powell,J.I., Yang,L., Marti,G.E., Moore,T., Hudson,J.Jr, Lu,L., Lewis,D.B., Tibshirani,R., Sherlock,G., Chan,W.C., Greiner,T.C., Weisenburger,D.D., Armitage,J.O., Warnke,R., Levy,R., Wilson,E., Grever,M.R., Byrd,J.C., Botstein,D., Brown,P.O. and Staudt, L.M. *Nature*, 2000, *403*, 503.

Austin,G.E., Alvarado,C.S., Austin,E.D., Hakami,N., Zhao,W.G., Chauvenet,A., Borowitz,M.J., and Carroll,A.J. *Am. J. Clin. Pathol.*, 1998, *110*, 575.

Ben-Dor,A., Bruhn,L., Friedman,N., Nachman,I., Schummer,M., Yakhini,Z. and Ben-Dor,A. In *Proceedings of the Fourth International Conference on Computational Molecular Biology (RECOMB2000)*, ACM press, New York, 2000.

Bittner,M., Meitzer,P., Chen,Y., Jiang,Y., Seftor,E., Hendrix,M., Radmacher,M., Simon,R., Yakhini,Z., Ben-Dor,A., Sampas,N., Dougherty,E., Wang,E., Marincola,F., Gooden,C., Lueders,J., Glatfelter,A., Pollock,P., Carpten,J., Gillanders,E., Leja,D., Dietrich,K., Beaudry,C., Berens,M., Alberts,D., Sondak,V., Hayward,N. and Trent,J. *Nature*, 2000, *406*, 536.

Brown,M.H., Cantrell,D.A., Brattsand,G., Crumpton,M.J., and Gullberg,M. *Nature*, 1989, *339*, 551.

Bull,J.H., Ellison,G., Patel,A., Muir,G., Walker,M., Underwood,M., Khan,F., and Paskins,L. *Br. J. Cancer*, 2001, *84*, 1512.

Cortes,C. and Vapnik,V. *Machine Learning*, 1995, *20*, 273.

Eisen,M.B., Spellman,P.T., Brown,P.O., and Botstein, D. *Proc. Natl. Acad. Sci. USA*, 1998, *95*, 14863.

Golub,T.R., Slonim,D.K., Tamayo,P., Huard,C., Gaasenbeek,M., Mesirov,J.P., Coller,H, Loh,M.L., Downing,J.R., Caligiuri,M.A., Bloomfield,C.D., and Lander,E.S. *Science*, 1999, *286*, 531.

Gordon,M.S., Kato,R.M., Lansigan,F., Thompson,A.A., Wall,R., and Rawlings,D.J. *Proc. Natl. Acad. Sci. USA*, 2000, *97*, 5504.

Greiner,A., Muller,K.B., Hess,J., Pfeffer,K., Muller-Hermelink,H.K., and Wirth,T. *Am. J. Pathol.*, 2000, *56*, 501.

Hombach,J., Lottspeich,F., and Reth,M. *Eur. J. Immunol.*, 1990, *20*, 2795.

Hombach,J., Tsubata,T., Leclercq,L., Stappert,H. and Reth,M. *Nature*, 1990, *343*, 760.

Judson,R. Genetic algorthms and their use in chemistry. *In* Lipkowitz,K.B. and Boyd,D.B. (eds), *Reviews in Computational Chemistry,* VCH publishers, New York, 1997, vol 10, pp 1-66,

Li,L., Darden,T.A., Weinberg,C.R. and Pedersen,L.G. *Combinatorial Chemistry & High Throughput Screening., in press.*

Li,L., Pedersen, L.G., Darden, T.A. and Weinberg, C.R. *Bioinformatics, in press.*

Malek,S.N., Dordai,D.I., Reim,J., Dintzis,H., and Desiderio,S. *Proc. Natl. Sci. Acad. USA*, 1998, *95*, 7351.

Manabe,A., Mori,T., Ebihara,Y., Koyama,T., Okuyama,I., Hosoya,R., Kaneko,M., Ishimoto,K., Nakahata,T., and Nakazawa,S. *Inter. J. Hematol.*, 1998, *67*, 45.

Massart,D.L., Vandeginste,B.G.M., Deming,S.N., Michotte,Y., and Kaufman, L. In *Chemometrics: a textbook (Data Handling in Science and Technology, vol 2)*; Elsevier Science B. V: New York, 1988, pp. 339-368.

Matthias P. *Semin. Immunol.*, 1998, *10*, 155.

Pekarsky,Y., Koval,A., Hallas,C., Bichi,R., Tresini,M., Malstrom,S., Russo,G., Tsichlis,P., and Croce,C.M. *Proc. Natl. Acad. Sci. USA*, 2000, *97*, 3028.

Perou,C.M., Sørlie,T., Eisen,M.B., van de Rijn,M., Jeffrey,S.S., Rees,C.A., Pollack,J.R., Ross,D.T., Johnsen,H., Aksien,L.A., Fluge,O., Pergamenschikov,A., Williams,C., Zhu,S.X., Lonning,P.E., Børresen-Dale,A.L., Brown,P.O. and Botstein,D. *Nature*, 2000, *406*, 747.

Ross,D.T., Scherf,U., Eisen,M.B., Perou,C.M., Rees,C., Spellman,P., Iyer,V., Jeffrey,S.S., Van de Rijn,M., Waltham,M., Pergamenschikov,A., Lee,J.C., Lashkari,D., Shalon,D., Myers,T.G., Weinstein,J.N., Botstein,D., and Brown,P.O. *Nat. Genet.*, 2000, 24, 227.

Schubart,D.B., Rolink,A., Kosco-Vilbois,M.H., Botteri,F., and Matthias,P. *Nature*, 1996, *383*, 538.

Scheuermann,R.H. and Racila,E. *Leuk. Lymphoma*, 1995, *18*, 385.

Sterneck,E., Paylor,R., Jackson-Lewis,V., Libbey,M., Przedborski,S., Tessarollo,L., Crawley,J.N., and Johnson,P.F. Proc. *Natl. Sci. Acad. USA*, 1998, *95*, 10908.

Toronen.P, Kolehmainen.M, Wong.C, and Castren.E. *FEBS lett.*, 1999, *451*, 142.

Winterbourn,C.C., Vissers,M.C.M., and Kettle,A.J. *Curr. Opin. Hematol.*, 2000, *7*, 53.

Vandeginste,B.G.M., Massart,D.L., Buydens,L.M.C., De Jong,S., Lewi,P.J. and Smeyers-Verbeke,J. In *Handbook of Chemometrics and Qualimetrics, Part B,* Elsevier Science, The Netherlands, 1998.

Virgilio,L., Isobe,M., Narducci,M.G., Carotenuto,P., Camerini,B., Kurosawa,N., Abbas-ar-Rushdi, Croce,C.M., and Russo,G. *Proc. Natl. Acad. Sci. USA*, 1993, *90*, 9275.

7

CLASSICAL STATISTICAL APPROACHES TO MOLECULAR CLASSIFICATION OF CANCER FROM GENE EXPRESSION PROFILING

Jun Lu[1], Sarah Hardy[1], Wen-Li Tao[1], Spencer Muse[1], Bruce Weir[1] and Susan Spruill[2]

[1]Program in Bioinformatics, NC State University and [2]PPGx. Inc., a subsidiary of DNA Sciences.

Abstract: Recent literature regarding microarray technology has focused on the need to incorporate classical statistical practices in experimental design in order to utilize more robust, classical statistical methodologies in data analysis. We have demonstrated that classical statistical methods are applicable to analysis of data previously presented by Golub, et al. 1999. Our preliminary analysis of all 6817 genes involves simple t-tests for statistically significant separation of means of gene expression level in two cancer types. Our subsets of genes that distinguish AML types from ALL types are relatively consistent with those published by Golub. We select those predictor genes based on the t-values and stepwise discriminant analysis, and evaluate the resulting model's performance in predicting 34 test samples by linear discriminant analysis. Only two samples were not correctly predicted (samples 61 and 66) with 25 predictor genes we chose. We also evaluate the parsimony of our model by evaluating, through a stepwise method, the minimum number of genes required to maintain a high level of accuracy in predicting cancer types.

Key words: discriminant analysis, stepwise, leukemia, SAS PROC DISCRIM

INTRODUCTION

In a Nature Genetics article Bittner, et al. [Bittner, 1999] acknowledged that the volume of data obtained from gene expression analysis using microarray presented a "mathematical challenge". This followed an earlier argument by Duggan et al. [Duggan, 1999] that in order for microarray research to achieve true understanding of genome function, it needed to recruit the assistance of statisticians and mathematicians to ponder the problems of data analysis. Since these publications were released more than 500 peer-reviewed articles referencing microarray technology have been published, and the number is rapidly climbing. Yet only 36 articles are currently available through PUBMED that seriously address the use of robust statistical methodology to analyze gene expression data.

While microarray technology is somewhat novel, the statistical methods appropriate for analyzing expression data need not be. Recent literature has focused on the need to incorporate classical statistical practices in experimental design in order to utilize more robust, classical statistical methodologies in the analysis of microarray data. We should clarify that for the purposes of this article, our use of the term "classical statistics" is meant to embody any statistical methodology that has been proven and accepted for use in other areas of science. Examples of the use of well-established statistical methodology can be found in articles such as Hilsenbeck, et al. [Hillsenbeck, 1999] who illustrate the use of principal components. Findings of Kaminski, et al. [Kaminski, 2000] are among the many emerging examples of the use of cluster analysis to distinguish cell samples with array expression data. The simple use of t-tests with corrections for multiple comparisons can be found in Dudoit, et al. [Dudoit, 2000], while Kerr and Churchill [Kerr, 2001] describe analysis of variance approaches.

Golub, et al. [Golub, 1999] are among the first to attempt to utilize expression results in a predictive modeling process. The objectives of Golub, et al. are to use a quantitative process to identify discriminating genes and then develop a predictive model for classifying samples as coming from patients with either acute lymphoblastic leukemia (ALL) or acute myeloid leukemia (AML). Golub, et al. present a molecular classification methodology for predicting tumor class of unknown samples. Their method utilizes a derived correlation coefficient in which the probability of a high density of genes distinguishes a class compared to a random pattern. From this "neighborhood analysis", the predictor is derived from the weight of each gene's "vote" in favor of one of the classes. The sum of these votes determines a prediction strength score, which in turn assigns the sample to a class based on a predetermined threshold.

While Golub and colleagues succeed in demonstrating that their predictive model was successful in distinguishing samples, their approach was a novel twist deviating from well-established statistical methodology. In addition, the final assignment of samples to class is based on an ambiguously determined threshold. The purpose of this article is to demonstrate that results similar to Golub, et al. are obtainable by employing analyses available in commercial software. We will illustrate the use of standard t-tests and stepwise discriminant analysis as is available in SAS

(SAS Institute, Cary, NC) to identify predictor candidate genes, develop a predictive discriminant model and test that model for accuracy.

METHODS

Data manipulation

Golub's study consisted of two datasets: a training dataset containing 38 samples and a test dataset containing 34 samples. In an effort to allow for direct comparability to Golub's results, we maintained these datasets for training and testing in our analyses. Prior to analysis, data from both the training and tests dataset were filtered and standardized according to methods as described in detail on Golub's web site. Briefly, any genes with an "A" labeling call for all 72 samples were excluded from further analysis. For the remaining genes, intensity readings were re-scaled based on background to allow for comparisons among the different samples.

We chose to exclude all the endogenous control genes from our data analysis, because we assumed that they would be less likely to differentiate samples. Additionally, because readings less than 20 were considered by Golub to be below the reliability level of the instrumentation, we also excluded genes with readings less than 20 in all 38 training samples and all 34 testing samples. After these exclusions, 4892 genes remained. In the remaining set, any individual intensity value less than 20 was set to 20 to avoid having readings below the lower bound. The logarithms of the Golub's re-scaled values were then standardized to mean zero and unit variance across all 72 samples.

Variable (gene) selection

Our ultimate goal was to identify the set of genes from the field of 4892 candidates that could most successfully be used to distinguish between the two leukemia types (ALL and AML). To achieve this goal we employed a three step screening process. Our first step was to reduce this large field of candidate genes to a smaller set for which subsequent statistical analysis would be more computationally feasible. The initial screen was accomplished by computing individual two sample t-statistics to compare the difference in expression levels between the ALL (n=27) and AML (n=11) groups in the 38-sample training set. The test statistics were computed under the assumption of independent samples with potentially different variation in the gene expression. We selected the 250 genes with the most positive t-values and the 250 genes with the most negative t-values to be retained as candidates for further analysis. In other words, the 250 genes exhibiting the greatest differences that tended to have high expression in the ALL group but low expression in the AML group were chosen, and visa versa. This strategy was adopted so that an equal number of genes favoring each of the two leukemia types would be represented.

The second step was to perform a stepwise discriminant analysis using the SAS procedure PROC STEPDISC [SAS, 1990]. The stepwise discriminant analysis selects variables (genes) that contribute most to the discriminatory power of the model as measured by Wilks' lambda likelihood ratio criterion. The set of variables is assumed to be multivariate normal with a common covariance matrix. We used the default significance level entry criterion of 0.15. Both the forward selection method and the stepwise method resulted in a final model with the same 25 variables.

The third step in our gene selection process was to create incrementally smaller subsets of these 25 genes by setting the STOP option in PROC STEPDISC to 20, 15, 10, and 5. This resulted in 5 subsets of genes, each with 25, 20, 15, 10 or 5 genes respectively, to be tested and compared in the final analysis.

Discriminant analysis

The final discriminant models (one for each of the 5 candidate datasets) were constructed using the SAS procedure PROC DISCRIM. A nonparametric method was employed, as opposed to the parametric model inherent in the stepwise discriminant analysis. This method was chosen in order to avoid making the assumption of multivariate normal distribution. The nonparametric method does not rely on this assumption.

Presented here is a basic overview of the nonparametric discriminant algorithm. For more details the reader is referred to the SAS Statistical User's Guide [SAS, 1990]. First, a nonparametric multivariate probability density function is estimated for each group; in our case the two groups are the ALL and AML patients in the training dataset. The SAS software allows for different choices of kernel smoothers for generating the probability density functions, and we chose a normal smoothing kernel. The algorithm in this nonparametric method is simple; for each multidimensional observation, **x**, in a test dataset, the estimated densities, based on the training dataset, are calculated separately for each group. Then, for each sample, a posterior probability of group membership is calculated for each group based on prior probabilities, the previously discussed density estimates, and the unconditional or overall density estimate. However, in its simplest form, where the prior group probabilities are equal, the observation **x** is classified into the group with the highest density estimate at **x**. The POOL and CROSSVALIDATE options were also used in PROC DISCRIM. When POOL=yes, the generalized squared distances are calculated based on a pooled within-group covariance matrix. The CROSSVALIDATE option provides the cross-validation classification of the training dataset. Each sample in the training set is classified based on the information of estimated group-specific density from all observations in the training when the nonparametric method with normal kernel is specified.

RESULTS AND DISCUSSION

Selected t-test results appear in Table 1, which shows the 10 genes with largest t-values (highly expressed in AML group) and the 10 genes with the most negative t-values (highly expressed in ALL group). In genes that favored one group or the other, it was observed that the between group differences were largest when the genes favored AML.

Table 1. Results from T-Test

10 Genes where t-value favors AML				
Gene Accession Number	**Gene**	**Standardized Mean ALL**	**Standardized Mean AML**	**t-value***
M27891_at	Cystatin C	-0.50819	1.24738	-9.49089
U50136_rna1_at	LTC4S	-0.50909	1.24958	-8.48094
X95735_at	Zyxin	-0.43795	1.07497	-8.11403
M55150_at	FAH	-0.42285	1.03791	-7.74088
Y12670_at	LEPR	-0.46825	1.14935	-7.69674
M23197_at	CD33	-0.50289	1.23435	-7.32637
U46499_at	GSTM	-0.46931	1.15193	-7.23349
M27783_s_at	ELA2	-0.48877	1.19972	-7.00355
M81933_at	CDC25A	-0.40481	0.99363	-6.89141
D88422_at	Cystatin A	-0.48973	1.20207	-6.69879
10 Genes where t-value favors ALL				
X66533_at	GUCY Beta-1	0.34207	-0.83963	6.12532
X82240_rna1_at	TCL1	0.34809	-0.85439	6.28326
M89957_at	IGB	0.35314	-0.86681	6.41754
L13278_at	CRYZ	0.38265	-0.93924	6.45551
Z15115_at	TOP2B	0.41605	-1.02121	6.52462
M77142_at	Nucleolysin TIA-1	0.41304	-1.01383	6.56631
M28170_at	CD19	0.41224	-1.01187	7.39117
M31523_at	TCF3	0.4596	-1.12810	7.47356
U22376_cds2_s_at	C-myb	0.49309	-1.2103	7.78074
X52142_at	CTPS	0.40777	-1.00089	8.05098

The 25 genes obtained from stepwise discriminant analysis are listed in Table 2, along with the partial R-square, F value, and the p-value from the step where the gene first entered the model. The last column in the table is the average squared canonical correlation (ASCC), a measure of the degree of separation between the groups.

Table 2. Genes obtains from Stepwise Discriminant Analysis

Gene Accession Number	Gene	Partial R-Square	F Value	Pr > F	Average Squared Canonical Correlation
M84526_at	DF D component of complement (adipsin)	**0.6685**	**72.61**	**<.0001**	**0.6685**
Z14982_rna1_at	MHC-encoded proteasome subunit gene LAMP7-E1 gene	0.2677	13.16	0.0009	0.7893
X95735_at	Zyxin	**0.4835**	**33.7**	**<.0001**	**0.9050**
U25128_at	PTH2	0.2609	12.71	0.001	0.9337
M27891_at	Cystatin C	**0.651**	**67.16**	**<.0001**	**0.9476**
X03934_at	T-cell antigen receptor T3-delta	0.1336	5.55	0.024	0.9612
X64364_at	BSG	0.2706	13.36	0.0008	0.9699
X64072_s_at	Leukocyte adhesion protein beta subunit	0.317	16.71	0.0002	0.9768
M98399_s_at	CD36	0.4343	27.64	<.0001	0.9858
L08177_at	CMKBR7	0.5278	40.24	<.0001	0.9910
M62762_at	ATP6C	**0.2707**	**13.36**	**0.0008**	**0.9959**
D38524_at	NT5 5' nucleotidase (CD73)	0.276	13.72	0.0007	0.9969
L19872_at	AHR	0.3186	16.83	0.0002	0.9976
X16832_at	Cathepsin H	0.1674	7.24	0.0108	0.9981
D26579_at	Transmembrane protein	0.3624	20.46	<.0001	0.9987
M27783_s_at	ELA2	0.6022	54.51	<.0001	0.9990
Y00787_s_at	IL-8 Precursor	**0.5408**	**42.39**	**<.0001**	**0.9993**
X03363_s_at	ERBB2	0.1744	7.61	0.0091	0.9995
D79997_at	KIAA0175	0.1373	5.73	0.022	0.9997
S75256_s_at	HNL neutrophil lipocalin	0.2378	11.23	0.0019	0.9998
M12959_s_at	TCRA	0.3591	20.17	<.0001	0.9999
X62654_rna1_at	ME491	0.4282	26.96	<.0001	0.9999
M13792_at	ADA	**0.4184**	**25.9**	**<.0001**	**0.9999**
M95178_at	Alpha-Actinin 1	0.219	10.1	0.003	0.9999
Y00433_at	GPX1	0.2362	11.13	0.002	0.9999

After 25 variables were selected, no other variables met the entry criterion. With these 25 variables, the ASCC is .9999 indicating that the two groups can be completely distinguished. Furthermore, the reader should note that after 6 variables have entered the model the average squared canonical correlation is .9612 and each subsequent improvement is less than .01. Among 25 genes, eighteen had larger AML means and 7 genes had larger ALL means, which is consistent with the observation that more genes showed high expression in AML than in the ALL group. Figure 1 graphically illustrates the separation of genes. The diagonal reference line indicates where genes would fall if there was no distinction between leukemia types. Genes falling above the reference line have, on average, higher expression in AML samples, while those falling below the line have, on average, higher expression in ALL samples.

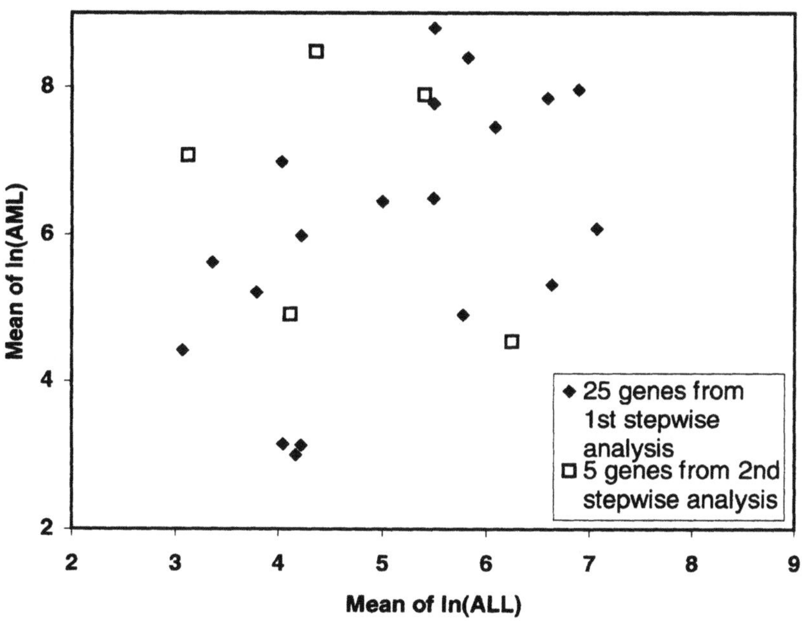

Figure 1. Scatter plot of means of AML and ALL for 25 genes and 5 genes used in discriminant analyses

How does this affect the analysis? When classification is done between two groups, if an observation does not fit the profile of one group, it is, by default assigned to the other group. We believe that in this analysis, AML was the primary group in the sense that the genes associated with AML dominated the models. The scientific implication of this in terms of using these genes to classify other types of cancers is not clear, but it should be an important consideration because it is highly likely that while these genes may successfully classify AML leukemia, they could fail at distinguishing between ALL and other types of leukemia for example.

To evaluate how successful the 25 genes are in determining the group membership we first conducted a cross-validation analysis on 38 training samples and then used the model based on the 38 training samples to predict group membership for the 34 testing samples. The cross-validation classification based on these 25 genes was accurate for all 38 training samples. The cross-validation results from the PROC DISCRIM procedure were somewhat optimistic since the group-density estimation was based on all observations in the training set when the nonparametric method with normal kernel was used. When the test data of 34 samples were classified based on density estimates from the training dataset, the classification outcomes on the test dataset were correct for 32 of the 34 samples (see Table 3, column 3). We next ask the question: can we further decrease the number of genes without decreasing the prediction accuracy? Recall that based on using the stepwise discriminant analysis with the STOP option we also selected sets of 20, 15, 10, and 5 genes. After assessing the effectiveness of the 25-gene model, we also, in the same manner, assessed the effectiveness of using the sets of 20, 15, 10, and 5 genes. These evaluations indicated that the 20, 15 and 10 gene models were equally successful in group classification, and the 5 gene model was actually slightly more effective, yielding 33 out of 34 correct classifications (see Table 3, column 4). This indicates that there is some degree of over-fitting occurring in the models with more genes. Of the five genes chosen (highlighted in **bold** in Table 2), one favored larger ALL means and the other 4 favored larger AML means (See Figure 1, squares). The only case that has been misclassified was sample 66, which was not correctly classified by Golub as well. It has been suggested that the problem may come from the experimental data on sample 66, which had weak hybridization signal for all the genes.

The five genes we finally chose were: Cystatin C (accession M27891), Adipsin (accession M84526), Zyxin (accession X95735), PTH2 parathyroid hormone receptor (accession U25128), and proteasome subunit LMP7 (accession Z14982). The first three genes were also in Golub's gene list. All three of these genes are more highly expressed in AML group than in ALL group samples.

The reader should note that the genes chosen by stepwise discriminant analysis are not necessarily those genes with highest absolute t-values. This is not a surprising result; we would expect to see this only if there was no multicollinearity amongst the genes with highest absolute t-values. Because the variable selection, based on discriminant analyses, takes into account the correlation between genes, it results in a set of genes that collectively account for differences between the groups. If only t-tests were used to select genes, two genes that are highly correlated could easily both be selected. When two genes are highly correlated, adding the second to the model in the presence of the first will not provide any additional improvement because it is supplying redundant information.

Table 3. Classification Results from Discriminant Analysis

Test Sample	Actual Group Classification	25 Gene Model Classification	5 Gene Model Classification
39	ALL	ALL	ALL
40	ALL	ALL	ALL
41	ALL	ALL	ALL
42	ALL	ALL	ALL
43	ALL	ALL	ALL
44	ALL	ALL	ALL
45	ALL	ALL	ALL
46	ALL	ALL	ALL
47	ALL	ALL	ALL
48	ALL	ALL	ALL
49	ALL	ALL	ALL
50	AML	AML	AML
51	AML	AML	AML
52	AML	AML	AML
53	AML	AML	AML
54	AML	AML	AML
55	ALL	ALL	ALL
56	ALL	ALL	ALL
57	AML	AML	AML
58	AML	AML	AML
59	ALL	ALL	ALL
60	AML	AML	AML
61	AML	***ALL***	AML
62	AML	AML	AML
63	AML	AML	AML
64	AML	AML	AML
65	AML	AML	AML
66	AML	***ALL***	***ALL***
67	ALL	ALL	ALL
68	ALL	ALL	ALL
69	ALL	ALL	ALL
70	ALL	ALL	ALL
71	ALL	ALL	ALL
72	ALL	ALL	ALL

We did verify that the model we selected was vastly superior to one based only on genes with significant t-test results as is illustrated by the fact that so few genes identified in Table 1 survived the stepwise discriminant analysis. This indicates that the genes are indeed correlated. When there is a high degree of correlation, as exists here, there can easily be other sets of variables that can predict as well as the specific sets chosen by the methods we implemented.

CONCLUSIONS

Established statistical methods, found in commercial software, were applied to classify samples from two leukemia types with better prediction outcomes than the methods used by Golub, et al. It is notable that only 3 of the 5 genes used in the reduced model are the same as those identified by Golub, suggesting a high level of redundancy among levels of gene expression within the leukemia types.

Perhaps the message to take away from this exercise is that it is not always necessary or prudent to develop novel analytical methods in tandem with novel technology. When a new problem is addressed with a new analysis method, the validity of the technology and the analysis are somewhat confounded. In fact, there is a certain sense of comfort that comes from successfully applying proven analytical methods to new problems. The very foundation of statistics is based on answering the basic question: Is what we observe in the experiment any different from what we would expect to see by random chance? When we lose sight of this basic question, we overcomplicate the problem and then look for complex solutions. In doing so, we fail to recognize that the problem is analogous to other previously solved problems and thus we fail to consider the applicability of standardized analysis methods. For example, the problem of distinguishing tumor type based on gene expression levels is analogous to distinguishing crop types based on remote sensing data (Example 5 in SAS User's Guide). By recognizing this similarity, we can more readily see that discriminant analysis was worth consideration for analyzing the tumor data. We stand to gain a lot by looking to the past for methods to address the problems of the future.

ACKNOWLEDGEMENTS

This work was supported in part by grants from NIH GM45344 and NSF 99872631.

REFERENCES

Bittner, Michael, Meltzer, Paul, and Trent, Jeffrey, 1999, Data analysis and integration: of steps and arrows. Nature Genetics. Vol 22, pp213-215.

Dudoit, Sandrine, Yang, Yee Hwa, Callow Matthew J., and Speed, Terence P., 2000, Statistical methods for identifying differentially expressed genes in replicated cDNA microarray experiments. Technical Report #578, www.stat.Berkeley.EDU/users/terry/zarray/html/matt.html

Duggan, David J., Bittner, Michael, Chen, Yidong, Meltzer Paul, and Trent, Jeffrey M., 1999, Expression profiling using cDNA microarrays. Nature Genetics, Vol 21, pp10-14.

Golub, T.R., Slonim, D.K., Tamayo, P., Huard, C., Gaasenbeek, M., Mesirov, J.P., Coller, H. Loh, M.L., Downing, J.R., Caligiuri, M.A., Bloomfield, C.D. and Lander, E.S., 1999. Molecular classification of cancer: Class discovery and class prediction by gene expression monitoring. Science, Vol 286, pp531-537.

Golub's web site (www.genome.wi.mit.edu/MPR)

Hilsenbeck, Susan G., Friedrichs, William E., Schiff, Rachel, O'Connell, Peter, Hansen, Rhonda K., Osborne, Kent, and Fuqua, Suzanne A.W., 1999, Statistical analysis of array expression data as applied to the problem of tamoxifen resistance. J Nat. Cancer Inst, Vol 91.5, pp453-459

Kaminski, Naftali, Allard, John D., Pittet, Jean F., Zuo, Fengrong, Griffiths, Mark J.D., Morris, David, Huang, Xiaozhu, Sheppard, Dean, and Heller, Renu A., 2000, Global analysis of gene expression in pulmonary fibrosis reveals distinct programs regulating lung inflammation and fibrosis. PNAS, Vol 97.4,pp1778-1783

Kerr, M.K. and Churchill, G.A., 2001, Statistical design and the analysis of gene expression microarray data. Genet. Res. Apr: 77(2), pp 123-128.

PUBMED (http://www.ncbi.nlm.nih.gov/entrez/query.fcgi)

SAS/STAT User's Guide (V6.04), 1990. SAS Institute, Inc., Cary, NC, USA

8

CLASSIFICATION OF ACUTE LEUKEMIA BASED ON DNA MICROARRAY GENE EXPRESSIONS USING PARTIAL LEAST SQUARES

Danh V. Nguyen[1] and David M. Rocke[1,2]
[1]Center for Image Processing and Integrated Computing and [2]Department of Applied Science, University of California, Davis, CA 95616

Abstract: Analysis of microarray data, when presented with raw gene expression intensity data, often take two main steps when analyzing the data. First pre-process the data by rescaling and standardizing so that overall intensities for each array are equivalent. Second, apply statistical methodologies to answer scientific questions of interest. In this paper, for the data pre-processing step, we introduce a thresholding algorithm for rescaling each array. Step 2 involves statistical classification and dimension reduction methodologies. For this we introduce the method of partial least squares (PLS) and apply it to the leukemia microarray data set of Golub et al. (1999). We also discuss the use of principal components analysis (PCA), quadratic discriminant analysis (QDA) and logistic discrimination (LD). Finally, we discuss other potential applications of PLS in analyzing gene expression data that address prediction of a target gene, prediction of the reaction in cell lines, assessment of patient survival, and generalisations in predicting multiple classes.

Key words: Dimension reduction, Logistic Discrimination, Prediction, Quadratic discriminant analysis

INTRODUCTION

DNA microarray technology has revolutionised biological and medicinal research. The use of DNA microarrays allows simultaneous monitoring of the expressions of thousands of genes. In a short period of time researchers have gathered a wealth of gene expression data from microarrays (such as high density oligonucleotide arrays and cDNA arrays). Prediction, classification, and clustering techniques are being used for analysis and interpretation of the data. For instance, molecular classification of acute leukemia [Golub et al., 1999], cluster analysis of tumor and normal colon tissues [Alon et al., 1999], clustering and classification of human cancer cell lines [Ross et al., 2000] and diffuse large B-cell lymphoma (DLBCL) [Alizadeh et al., 2000], human mammary epithelial cells and breast cancer [Perou et al., 1999, 2000], and skin cancer melanoma [Bittner et al., 2000] are some examples. These techniques have also helped to identify previously undetected subtypes of cancer [Golub et al., 1999; Alizadeh et al., 2000; Bittner et al., 2000; Perou et al., 2000]. The problem of "prediction" may come in various forms of applications as well; the prediction of patient survival duration with germinal centre B-like DLBCL compared to those with activated B-like DLBCL (gene expression subgroups) using Kaplan-Meier survival curves [Ross et al., 2000] or the prediction of a target gene expression are examples.

Gene expression data from DNA microarrays are characterized by many measured variables (genes) on only a few observations (experiments) although both the number of experiments and genes per experiment are growing rapidly. The number of genes on a single array are usually in the thousands, so the number of variables, p, easily exceeds the number of observations N. Although the number of measured genes is large there may only be a few underlying gene components that account for much of the data variation; for instance, only a few linear combinations of a subset of genes account for nearly all of the response variation. Under similar data structure in the field of chemometrics, the method of PLS (Partial Least Square) has been found to be very useful.

PLS has been useful as a predictive modelling regression method in the field of chemometrics. A typical example, in spectroscopy, is to predict the chemical composition of a compound based on observed signals for a particular wavelength, where the number of wavelengths (variables) is much larger than the number of available samples. Examples of PLS applications in chemometrics can be found in the *Journal of Chemometrics* (John Wiley) and *Chemometrics and Intelligent Laboratory Systems* (Elsevier). For an introduction to PLS regression the reader is referred to a tutorial paper by Geladi and Kowalski (1986). The use of PLS in calibration can be found in Martens and Naes (1989). Some theoretical aspects and data-analytical properties of PLS have been studied by chemometricians and statisticians [de Jong, 1993; Frank and Friedman, 1993; Goutis, 1996; Helland, 1988; Helland and Almoy, 1994; Hoskuldsson, 1988; Lorber et al., 1997; Phatak, et al., 1992; Stone and Brooks, 1990]. An interpretation of PLS based on sequences of simple linear regression can be found in Garwaithe (1994).

The method of PLS is also useful for various prediction problems based on gene expression data. We briefly introduce the method here and give details in the section on Methods. For simplicity we restrict to the case where the response variable Y is univariate. For instance, the response variable can be the expression values of a single "target" gene of interest, however, if one is interested in predicting the expression values of several genes (i.e., a "block" of genes) simultaneously PLS applies as well. Multiple linear regression (MLR) is perhaps the most popular method used to predict a response variable Y based on a set of predictor variables $X_1, X_2, \ldots, X_p$. It is well known that when there are more predictors than there are samples ($p > N$) the MLR model will fit the data perfectly, but the model will not predict new samples well. A common approach to deal with this problem is to first reduce the dimension of the data by constructing a few summary "components" and then use this reduced set of constructed components to predict the response Y. An example of this approach is principal components regression (PCR) [Massey, 1965; Jolliffe, 1986]. Here, principle component analysis (PCA) is used to reduce the high dimensional data to only a few gene components, which explain as much of the observed total gene expression variation as possible (subject to orthogonality and norming constraints). This is achieved without regards to the response (leukemia class) variation. Gene components constructed this way are called principal components (PCs).

In contrast to PCA, PLS components are chosen so that the sample covariance between the response and a linear combination of the p predictors is maximum. The latter criterion for PLS is more sensible since there is no a priori reason why constructed components having large predictor variation (gene expression variation) should be strongly related to the response variable (leukemia classes). Certainly a component with small predictor variance could be a better predictor of the response classes. The ability of the dimension reduction method to summarize the covariation between gene expressions and leukemia classes may yield better prediction results. We will demonstrate this contrast between PLS and PCA using the leukemia data.

This paper is organized was follows. In the Methods section we describe the thresholding algorithm for array rescaling and the details of the dimension reduction method of PLS and PCA. After rescaling and dimension reduction, we considered logistic and quadratic discrimination between the leukemia classes. Results and comparison to Golub et al. (1999) are given in the section on Results. We conclude and discuss the potential applications of PLS in gene expression analysis in the sections on Conclusions and Discussions respectively.

METHODS

A Thresholding Algorithm for Rescaling Gene Expressions

Pre-processing of gene array data is common for various reasons. The main reason is the non-uniform noises associated with array technologies observed from one array (sample) to another. Thus, array rescaling (or normalization) is important since subsequent analysis results are meaningful if the overall intensities for each array are equivalent. The use of thresholding is especially common in data pre-processing. For instance, discarding negative measurements (which occurs when a spot background intensity measurement exceeds the signal intensity) or very low signal to noise measurements. Although negative measurements (due to imperfect measurement technology) should not be used in the analysis of the genes, this information can be used to estimate the amount of noise present in each array.

In this section we describe a thresholding algorithm which finds a "cutoff" point for *each* array (hence, accounting for different levels of noise specific to a given arrays). Genes with expression levels below the cutoff point may be considered unreliable. Also, information from the algorithm is used to estimate the amount of noise present in a particular array. Gene expression levels are then rescaled so that overall intensities for each array are equivalent based on the estimate of the average noise for each array. The algorithm is derived from a two-component error model for gene expression arrays. Due to limited space, the reader is referred to Rocke and Durbin (2000) for details of this model.

We now describe the thresholding algorithm. Let the original gene expression values for the ith array be $x_1, x_2, \ldots, x_p$ and i=1, 2, ... , N is the number of arrays. For brevity of notation we denote the collection of expression values for array i by $\{x_j, j=1, p\}=\{x_{i1}, x_{i2}, \ldots, x_{ip}\}$ and assume that these values are sorted. The algorithm begins with an initial set, $A(0)$, consisting of noise values (negative measurements) and low signal to noise measurements (small intensity minus background). The median of $A(0)$, say m_0, gives an initial estimate of the amount of noise specific to the given array. Similarly, the usual sample standard deviation of $A(0)$, say s_0, gives an initial estimate of the dispersion of the noise measurements in the given array. Thus, expression values close to m_0 have high noise relative to signal intensity. The algorithm then updates the original noise set, $A(0)$, by adding to this set all expression values close to m_0. That is, enlarge the initial set consisting of noise measurements by adding to it expression measurements with high noise relative to signal intensity. How close to the initial estimate of array-specific noise, m_0, should a particular expression measurement be in order for it to be included in the noise set? One can, for instance, include all expression measurements within two (or three) standard deviations (s_0) of the initial array-specific noise estimate, m_0. Thus, expression values below a cutoff threshold of $u_0 = m_0 + c \cdot s_0$ would be added to the initial noise set $A(0)$, where c is the number of standard deviation away from the median noise estimate m_0. However, rather than using the usual sample standard deviation, which is highly effected by outliers, we used a robust measure of dispersion given by $s_0 = MAD_0 \ / \ 0.675$, where MAD_0=median $\{| \ x_j - m_0 \ |, j=1, n_0\}$ is the median absolute deviation about the median. Note that for the same reason we have also used the median, rather than the mean as a measure of location, to estimate the array-specific noise.

Thus, the updated noise set now consists of all the initial noise values in $A(0)$ and all expression values less than the cutoff threshold of $u_0 = m_0 + c \cdot s_0$. Call this enlarged noise set $A(1)$. A revised or updated estimate of the amount of noise present in the given array is the median of the updated noise set $A(1)$, denoted m_1. Similarly, the revised robust estimate of the dispersion of the updated noise set is $s_1 = MAD_1 \ / \ 0.675$. The updated noise set, $A(1)$, is again revised (updated) to include all expression values that are determined to be high in noise relative to signal intensity, namely all expression values less than $u_1 = m_1 + c \cdot s_1$. This updating process continues until the noise set $A(k)$ no longer changes. That is, the algorithm stops when the noise set $A(k)$ (for the ith array) converges. Denote this set by $A(n_i)$. Thus, at convergence, the set $A(n_i)$ consists of the smallest n_i values from array i. Convergence of the algorithm is not guaranteed and in rare instances we have observed that the algorithm fluctuates between two values. If these are not too different the midpoint can be taken, for instance.

Estimate of the mean or median array-specific noise can now be obtained by taking the sample mean or median of the set $A(n_i)$, for array i. (Any appropriate statistics based on $A(n_i)$ may be used.) Similarly, values in $A(n_i)$ can also be used to estimate the variation of noise within an array, hence providing a way for calculating a confidence interval for a gene expression value (when no replicated gene measurements are available). In most instances the algorithm converges in less than 15 iterations. The steps of the algorithm are given below.

Thresholding Algorithm Parameters: *q* and *c*

1. Select $q\%$ of the lowest expression values. Denote this initial set of values by $A(0) = \{x_j j=1, n_0\}$.
2. Calculate the median of the initial set, m_0= median$\{A(0)\}$.
3. Calculate the median of the absolute deviations about the median, MAD_0=median $\{|x_j - m_0|, j=1, n_0\}$ of the initial set of values $A(0)$.

4. Calculate the cutoff point, $u_0 = m_0 + c \cdot s_0$, where $s_0 = MAD_0/0.6745$ and c =2, 2.5, or 3 (is the number of median absolute deviation above the median).
5. Determine the new set defined by $A(1) = \{\text{all } x_j < u_0\}$.
6. Repeat steps 2 through 5 (for each new set $A(k)$) and stop when $n_k = n_{k-1}$ (convergence). At convergence denote the set of noise values by $A(n_I)$ (with size n_I) and the cutoff point by u_I for $I = 1, \ldots, N$.
7. Repeat steps 1 through 6 for each array, $I = 1, \ldots, N$.

As mentioned earlier, one application of the thresholding algorithm is to use the information from the algorithm to rescale the expression levels in each array so that overall intensities are equivalent across all arrays. We now describe this application. Suppose that the mean of $A(n_i)$, $m(i) = n_i^{-1} \Sigma j$ x_{ij} (i=1, …, N; j=1, …, n_i) is used to estimate the average noise from array i. One may consider a (1) multiplicative rescaling: $x_{ij} \leftarrow x_{ij}\ a_i$ or a (2) subtractive rescaling: $x_{ij} \leftarrow x_{ij} - b_i$, where $a_i = M(i)/m(i)$, $b_i = m(i)$ and $M(i)$ is the overall mean of the ith array. Using strategy (1) to rescale an array with high average noise would result in smaller (rescaled) expression values relative to expression values of another array with lower average noise (and similar overall average expression). Even baseline or control arrays are susceptible to errors since measurements come from the same system; hence, the algorithm can be applied there as well. We have suggested only two obvious strategies for rescaling the gene expression values and we focus on (2) in this paper. Comparisons to strategy (1) or to other types of transformations based on information from the algorithm can be examined in future studies.

As discussed above, the information from the algorithm used to rescale each array is given by the set of values $A(n_i)$, i=1, 2, …, 72, obtained from the algorithm. For example, strategy (1) and (2) described above are two ways of using the information in $A(n_i)$ to rescale each array. However, to obtain $A(n_i)$ (i.e., to start the algorithm) the user needs to specify two parameters, q (the starting value) and c (the number of standard deviations away from the median noise estimate).

Some natural questions arise regarding the parameters (q and c) of the algorithm. For instance, one may specify that 10% (q=10) of the expression values of the ith array be used to form the initial set A_0. Will the noise set at convergence, $A(n_i)$, be the same if we change q, i.e., change the initial set $A(0)$? The set $A(n_i)$ at convergence is insensitive to the starting value q when applied to the leukemia data. Golub et al. considered molecular classification of acute leukemia based on a 38 sample training data set and a 34 sample test data set. Samples were obtained from bone marrow and peripheral blood of acute leukemia patients. RNA was hybridized to high-density oligonucleotide microarrays (Affymetrix) with probes for 6,817 human genes. The thresholding algorithm was applied to each of the 72 arrays with different starting percentages (q) of 1%, 5%, 10%, 20% and 30% (with c=3). The resulting cutoff points at convergence were the same (for the various q) and only a few differ by negligible amounts.

An implicit assumption in developing the thresholding algorithm is that small expression values are the noise values; however, "small" is relative to the array. That is, the noise level is array-specific. The question is how small is small (for each array)? The answer is the cutoff u at convergence, which separates noise values from "real" expressed values. This depends on the parameter c, the number of median absolute deviation above the median. Increasing c corresponds to a more stringent standard, since expression values must be larger to be excluded from the noise set. Since the resulting cutoff point does not depend on q, we set q=10% and ran the thresholding algorithm for c=2.5 and 3. The results are given in Figure 1. The pattern of u_i for c=3.0 and u_i for c=2.5 is similar (Figure 1). Also evident from Figure 1 is that the estimate of array-specific noise varies a lot from one array to another. Therefore, it may not be optimal to use a single threshold value across *all* arrays and the thresholding algorithm avoids this.

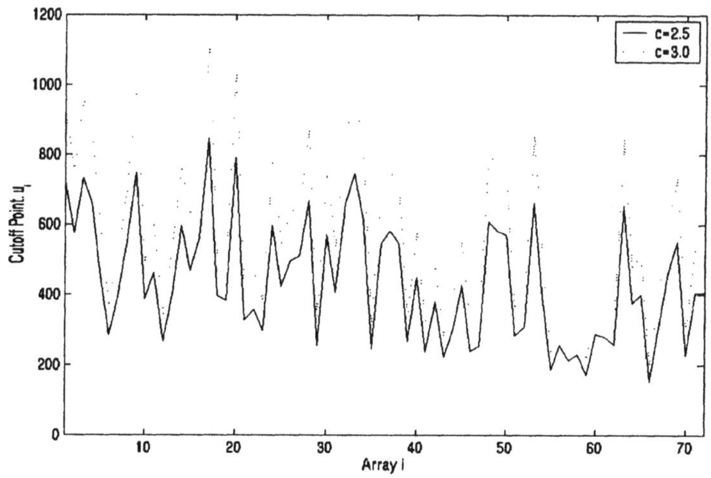

Figure 1. Cut off point for c = 2.0 and 3.0.

Although the example given here consists of high-density oligonucleotide arrays, the thresholding algorithm can be applied to cDNA arrays as well. Assume that after background subtraction we have intensity measurements for the red-fluorescent dye Cy5 and another for the green-fluorescent dye Cy3 for *i*th array. One strategy is to apply the above procedure to each set of dye measurements separately. After separate rescaling based on separate noise estimates for each channel, one can proceed to analyze the log(Cy5/Cy3) (positive) measurements. The reason for the separate applications of the thresholding algorithm to the sets of measurements from different channels is that the level of noise may be channel-specific.

DIMENSION REDUCTION: PARTIAL LEAST SQUARES

After array rescaling we examined dimension reduction methods to reduce the high p-dimensional gene space to a lower dimensional gene component space which can predict leukemia classes well. Some aspects of PLS are similar to principal component analysis (PCA), so we briefly review PCA. PCA is well known and is a commonly used dimension reduction method. Let $\mathbf{X}$ denote a data matrix of N rows, where each row consists of p gene expression values from a particular microarray experiment. (Thus, there are N samples or experiments.) Since the data dimension p is too large for many conventional statistical tools to be applied, PCA attempts to reduce this high gene dimension, p, to a much lower dimension, say K ($K < N$). This is achieved by extracting K gene components which are linear combinations of the original p genes. The components are extracted sequentially by maximizing the objective criterion, var($\mathbf{Xc}$), subject to orthogonality and norming constraints on the unit weight vector $\mathbf{c}$. Very roughly, the constructed components summarized as much of the information (variation) of the original p genes, irrespective of response (leukemia) class information.

Note that maximizing the *variance* of the linear combination of the genes, namely var($\mathbf{Xc}$), may not necessarily yield components predictive of leukemia classes. For this reason, a different objective criterion for dimension reduction (i.e., for selecting gene components) may be more appropriate for prediction of leukemia classes, for instance. The criterion for selecting gene components in PLS is to sequentially maximize the *covariance* between the leukemia classes and a linear combination of the p genes (also subject to orthogonality and norming constraints). That is, find the gene component weights, $\mathbf{w}$, such that cov($\mathbf{Xw}$, $\mathbf{y}$) is maximum, where $\mathbf{y}$ is the response vector of leukemia classes. We note that in PCA, the leukemia class information ($\mathbf{y}$) was not used in constructing the gene components.

Based on the different objective criterion of PCA and PLS (namely var($\mathbf{Xc}$) and cov($\mathbf{Xw}$, $\mathbf{y}$)) it is reasonable to suspect that *if* the original p

genes are predictive of the leukemia classes, then the constructed components from PCA would likely be good predictors of leukemia classes. Therefore, prediction results should be similar to that based on PLS gene components. Otherwise, PLS should perform better than PCA in predicting leukemia classes. The results based on the leukemia data given in the next section will support that this is indeed the case.

After dimension reduction by PLS (or PCA), the high dimension of p is reduced to a lower dimension of K gene components. We constructed K=3 gene components and predicted the leukemia classes based on the 3 components. (One can choose K, for example, by cross-validation. This was done but the final results did not change much so we choose the simpler model with K=3.) Since the reduced gene dimension is now low, conventional classification methods such as logistic discrimination and QDA (Quadratic Discriminant Analysis) can be applied. For an introduction to classical discrimination and classification the reader is referred to Flury (1997), Johnson and Wichern (1993), Mardia et al. (1979), and Press (1982). Details of LD (Logistic Discrimination) using PLS components including simulation results can be found in [Nguyen and Rocke, 2001].

RESULTS

Classification Based on 50 Predictive Genes

Although PLS can handle the number of genes, p, as large as expected in the human genome, experience indicates that prediction results are much improved when a subset of predictive genes are selected for the actual prediction or classification [Nguyen and Rocke, 2001]. This approach is reasonable since only a subset of the genes (probes) deposited on the array for investigation is predictive of the biological classes. For this reason and also for comparison purposes we first analyse the same p=50 genes obtained by Golub et al. (1999).

We constructed 3 gene components using PLS and PCA based on p=50 genes. Classification of the leukemia classes using the constructed gene components in logistic discrimination and quadratic discriminant analysis are compared. In this analysis we used the same set of 50 genes reported by Golub et al. as predictive of acute leukemia classes. That is, Golub et al. defined $P(G, Y) = (m_1 - m_2)/(s_1 + s_2)$ as a measure of "correlation" (or "distance") between gene G_j and the class indicator variable Y, where m_k and s_k, k = 1, 2, are the sample means and standard deviations of the log of the expression levels of gene j in class 1 and 2 respectively. Twenty-five genes closest to class 1 and another twenty-five genes closest to class 2 were selected to form a set of 50 "informative" genes. These are the selected predictors. Based on these 50 genes we carried out the following analysis steps:

1. Apply the thresholding algorithm to each array for (multiplicative) rescaling.
2. After rescaling, select the same 50 genes as in Golub et al.
3. Construct PLS components and PCs based *only* on the original 38 training samples.
4. Construct test components based on training component information.
5. Predict leukemia classes by (1) LD and (2) QDA using leave-one-out cross-validation (CV) for the 38 training samples and then make out-of-sample predictions for the 34 test samples based on the constructed components from training information.

Leave-one-out CV predictions of the 38 training samples using QDA and LD with PLS gene components resulted in 100% correct and 36/38 for PCs. Based only on the training components, out-of-sample predictions for the 34 test samples were made. LD with PLS gene components resulted in one misclassification. Five test samples with low "prediction strength" (PS, defined in Golub et al.), samples # 54, 57, 60, 67, and 71 (with PS=0.23, 0.22, 0.06, 0.15, and 0.30), and one misclassified by Golub et al. were correctly classified with estimated conditional class probabilities of 0.97, 1.00, 0.98, 0.89, and 1.00. The second misclassified sample (# 66) by Golub et al. was the single misclassification by LD using PLS gene components. (This was the same sample misclassified by all CAMDA'00 conference participants.) LD using PCs did not predict the test sample well (with 7 incorrect). However, QDA using PLS components or PCs both had three misclassifications in the test samples.

Figures 2(a)-(d) shows how the three constructed PLS gene components and PCs separate the leukemia classes in the training and test data. It can be seen that PLS components separate the leukemia classes better than PCs.

Assessment of Constructed Components for Classification

To assess the performances of (1) the various component construction methods and (2) the classification methods (LD and QDA), re-randomization is necessary. (Re-randomization was not considered by Golub et al. so a comparison here is not possible.) We considered a re-randomization scheme with equal splitting of the 72 samples into 36 training samples and the remaining 36 into test samples. The analyses above were repeated for 50 re-randomizations and the results indicate that the prediction based on the 50

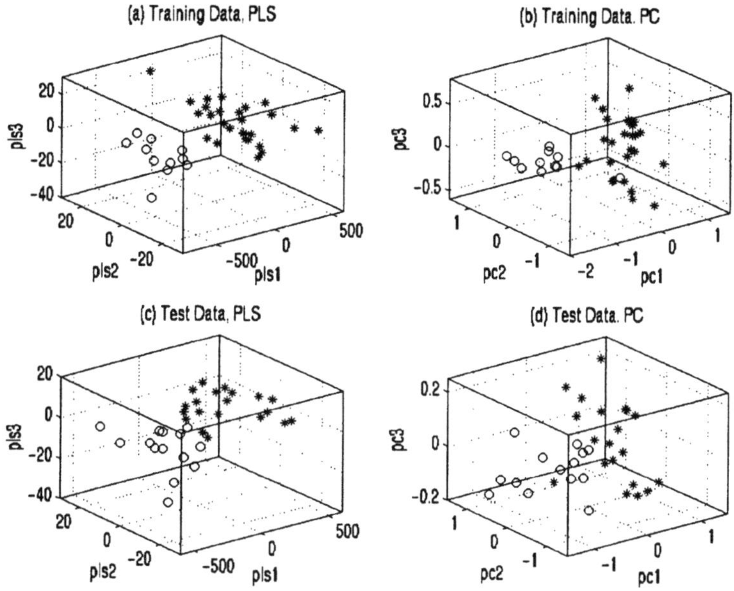

Figure 2. Separability of leukemia samples in the training and test data sets by three gene components extracted by PLS and PCA.

genes is stable--it is unlikely that the results depend on the original data configuration of 38 training/34 test samples split. The average percentages of correct classification over 50 re-randomizations are given in Table 1. Note that for each of the 50 randomizations (data sets), the prediction for the 36 training samples is based on leave-one-out CV and prediction for the remaining 36 test samples are based on out-of-sample prediction. Based on the 50 re-randomizations, PLS gene components performed at least as well or better than PCs on the training data for all (50/50) re-randomizations. This is based on leave-one-out CV. For the out-of-sample prediction of the test samples, PLS predicted at least as well or better than PCs in 42/50 re-randomizations.

Table 1. Average percentage correct in 50 re-randomizations.

	LD	LD	QDA	QDA
	PLS	PC	PLS	PC
Training Data	99.56	96.44	99.56	97.00
Test Data	95.94	94.17	96.44	95.44

A Condition When PCs Fail to Predict Well

Attempts have been made to study the settings where PLS will predict well relative to other prediction methods, however, the conditions under which PLS predicts well have not been fully characterized in the statistics or chemometrics literature. The leukemia data set of Golub et al. provides an example of a condition under which PLS performs well relative to PCA. In the analyses given above, although the results for PLS components were better than that for PCs, the results for PCs were competitive nonetheless. Examining the objective criterion of PLS and PCA, we noted in the Methods section that it would be reasonable to expect prediction based on PCs to be similar to that from PLS if the original genes are highly predictive of leukemia classes. This is the case of the analyses based on the 50 predictive genes. However, to see when PCs fail to predict well, while PLS components succeeded, we consider their prediction ability based only on expressed genes, but not expressed differentially for leukemia classes. This test condition is based on the simple fact that an expressed gene does not necessarily qualify as a good predictor of leukemia classes. For instance, consider a gene highly expressed across all samples, ALL and AML. In this case, the gene will not discriminate between ALL and AML well.

We define a gene to be expressed if the measure expression value is above the threshold value u determined from the thresholding algorithm (with parameter q=10% and c=3.0) as described in the Methods section. After rescaling, subset of p genes were retained for analysis. We analyzed five (nested) sets of genes defined as expressed on (A) at least one array (p=1,554), (B) 25% (p=1,076), (C) 50% (p=864), (D) 75% (p=662) and (E) 100% (p=246) of the arrays. As before, we applied PLS and PCA to extract gene components from these five data sets based on the 38 training samples. Predictions of the 38 training samples were based on leave-one-out CV and predictions of the 34 test samples were based on the training components only. This leads to a drastic decrease in performance of PCs relative to PLS. (Results not given here. For details, see Nguyen and Rocke (2001).)

To assess whether the observed decrease in performance of PCs relative to PLS gene components is coincidental, we considered a re-randomization study as in the analysis of the 50 informative genes. PCs did much worse relative to PLS gene components in the re-randomizations as well. The result here is not surprising since PCA aims to summarize only the variation of the p genes. These p genes are above the threshold and therefore considered expressed. However, only a subset of p expressed genes are predictive of leukemia classes. Why then does PLS components still perform well in this mixture of expressed genes, both predictive and non-predictive of leukemia classes? This is most likely attributed to the choice of objective criterion used, namely covariation between the leukemia classes and (the linear combination of) the *p* genes. Since PLS components are obtained from maximizing cov(**Xw**, **y**) it is more able to assign pattern of weights to the genes which are predictive of leukemia classes. Details of the re-randomization study can be found in Nguyen and Rocke (2001).

CONCLUSIONS

In this paper we have presented a thresholding algorithm for estimating the array-specific noise and outlined rescaling procedure based on information from the thresholding algorithm. The algorithm is robust to outlying observations and is not affected by starting values (q). For example, we have applied the algorithm to the leukemia data set of Golub et al. We also described the use of the thresholding algorithm for cDNA arrays. After array rescaling, we introduced the use of PLS as a dimension reduction method for the analysis of gene expression data and illustrated the method's effectiveness in predicting leukemia classes. Specifically, we have provided examples on the use of PLS gene components for classifying leukemia classes by quadratic and logistic discriminant analysis. The classification results are favorable when compared to the original results of Golub et al., as well as those based on PCs. These results hold under re-randomization studies as well. Finally, we have described a condition under which PLS components are superior to PCs for prediction purposes.

DISCUSSION

Easy Data Set?

It is now well known (CAMDA'00 conference) that the leukemia data set is an easy data set. That is, the leukemia classes are easily separable with the exception of one sample in the test data set. Nonetheless, the data set exhibits interesting characteristics that led to an elucidation of PLS and PCA for the analysis of microarray data. However, we have found similar successes with PLS in other data sets as well. Additional examples of cancer classifications, including normal versus ovarian tumor, diffuse large B-cell lymphoma versus B-cell chronic lymphocytic leukemia, normal versus colon tumor, and non-small-cell-lung-carcinoma versus renal samples, using PLS gene components can be found in Nguyen and Rocke (2001). Furthermore, the method performs well under a simulation model for gene expression data [Nguyen and Rocke 2000, 2001b, 2001c].

Potential Applications of PLS to Microarray Data

Much published work in the analysis of gene expression data focuses on clustering genes. One motivation for the use of clustering is that "similar" gene expressions that are clustered together have similar functions, hence providing clues to gene functions. In addition to the wide use of clustering, we will discuss some other applications for microarray data in this section. Another area of interest (currently not as popular as clustering genes) may be the prediction of the expression of a target gene. Quantifying the predicted gene expression values such that they are compatible with some clinical outcomes is one use of this PLS prediction. Another use involves gene expressions measured over time. PLS prediction can be used to predict a target gene (or a cluster of target genes) over time. The goal of such an analysis may be to see which genes are related to the target gene and how this relationship varies with time.

Assessing the relationship between cellular reaction to drug therapy and their gene expression pattern is another application. For example, Scherf et al. (2000) assessed growth inhibition from tracking changes in total cellular protein (in cell lines) after drug treatment. The response of cell lines to each drug treatment can be considered as the response variables. Associated with the cell lines are their gene expressions. Since the expression patterns are from those of untreated cell lines, Scherf et al. focused on the relationship between gene expression patterns of the cell lines and their sensitivity to drug therapy. This relationship can be studied via a direct application of univariate or multivariate PLS, which can handle the high dimensionality of the data.

Another example, in cancer research, is the prediction of patient survival based on gene expressions. However, here some of the observed times are censored. Based on a simulation model of gene expression data, Nguyen and Rocke (2000c) studied various dimension reduction strategies involving PLS and PCA in the context of survival data with gene expressions as covariates. The results based on PLS are favorable based on this preliminary simulation study.

Finally, since many classification problems involve more than two classes, such as the classification of various types of tumors (leukemia, breast, renal, central nervous system, etc.), generalization to more than two classes is needed. Preliminary study of multivariate PLS for predicting multiple classes under a simulation model for gene expression data indicates that the method is useful [Nguyen and Rocke, 2000b].

PLS Algorithm and Computing Feasibility

One clear advantage of PLS over other dimension reduction method, such as PCA, is its computational feasibility. PLS components can be constructed in a few seconds, despite the high dimension p. Other dimension reduction methods, such as PCA, are much more computationally intensive when p is in the thousands. The PLS algorithm can be found, for instance, in Helland (1988), Hoskuldsson (1988), or Martens and Naes (1989).

ACKNOWLEDGMENTS

The research reported in this paper was supported by grants from the National Science Foundation (ACI96-19020, and DMS 98-70172), the National Institute of Environmental Health Sciences, National Institutes of Health (P43ES04699) and the National Cancer Institute (CA-90301). We thank the editors for comments that improved the presentation and content of the paper.

REFERENCES

Alon et al. (1999), "Patterns of Gene Expression Revealed by Clustering Analysis of Tumor and Normal Colon Tissues Probed by Oligonucleotide Arrays," *Proceedings of the National Academy of Sciences*, 96, 6745-6750.

Alizadeh et al. (2000), "Distinct Types of Diffuse Large B--Cell Lymphoma Identified by Gene Expression Profiling," *Nature*, 403, 503-511.

Bittner et al. (2000), "Molecular Classification of Cutaneous Malignant Melanoma by Gene Expression Profiling," *Nature*, 406, 536-540.

de Jong, S. (1993), "SIMPLS: An Alternative Approach to Partial Least Squares Regression," *Chemometrics and Intelligent Laboratory Systems*, 18, 251-263.

Dudoit, S., Fridlyand, J., Speed, T.P. (2000), "Comparison of Discrimination Methods for the Classification of Tumors Using Gene Expression Data," Technical Report # 576, Department of Statistics, U. C. Berkeley.

Flury, B. (1997), *A First Course in Multivariate Analysis*. Springer-Verlag, New York.

Frank, I.E., and Friedman, J.H. (1993), "A Statistical View of Some Chemometric Regression Tools" (with discussion), *Technometrics*, 35, 109-148.

Garthwaite, P.H. (1994), "An Interpretation of Partial Least Squares," *Journal of the American Statistical Association*, 89, 122-127.

Geladi, P., and Kowalski, B.R. (1986), "Partial Least Squares Regression: Tutorial," *Analytica Chimica Acta*, 185, 1-17.

Golub et al. (1999), "Molecular Classification of Cancer: Class Discovery and Class Prediction by Gene Expression Monitoring," *Science*, 286, 531-537.

Hand, J.D. (1981), *Discrimination and Classification*. John Wiley Sons, Chichester, England.

Hand, J.D. (1997), *Construction and Assessment of Classification Rules*. John Wiley Sons, Chichester, England.

Helland, I.S. (1988), "On the Structure of Partial Least Squares," *Communications in Statistics-Simulation and Computation*, 17, 581-607.

Helland, S., and Almoy, T. (1994), "Comparison of Prediction Methods When Only a Few Components are Relevant," *Journal of the American Statistical Association*, 89, 583-591.

Hoskuldsson, A. (1988), "PLS Regression Methods," *Journal of Chemometrics*, 2, 211-228.

Johnson, R.A. and Wichern, D.W. (1992), *Applied Multivariate Analysis*. Prentice-Hall, New Jersey, 4th edition.

Jolliffe, I.T. (1986), *Principal Component Analysis*. Springer-Verlag, New York.

Lorber, A., Wangen, L.E., and Kowalski, B.R. (1997), "A Theoretical Foundation for the PLS Algorithm," *Journal of Chemometrics*, 1, 19-31.

Mardia, K.V., Kent, J.T., and Bibby, J..M. (1979), *Multivariate Analysis*. Academic Press, London.

Martens, H. and Naes, T. (1989), *Multivariate Calibration*, John Wiley Sons, New York.

Massey, W.F. (1965), "Principal Components Regression in Exploratory Statistical Research," *Journal of the American Statistical Association*, 60, 234-246.

Nguyen, D.V. and Rocke, D.M. (2000), "Classification in High Dimension with Application to DNA Microarray Data," manuscript.

Nguyen, D.V. and Rocke, D.M. (2001), "Tumor Classification by Partial Least Squares Using Microarray Gene Expression Data," to appear in *Bioinformatics*.

Nguyen, D.V. and Rocke, D.M. (2001b), "Partial Least Squares Proportional Hazard Regression for Application to DNA Microarray Data," manuscript.

Nguyen, D.V. and Rocke, D.M. (2001c), "Multi-Class Cancer Classification Via Partial Least Squares Using Gene Expression Profiles," manuscript.

Perou et al. (2000), "Molecular Portrait of Human Breast Tumors," *Nature*, 406, 747-752.

Perou et al. (1999), "Distinctive Gene Expression Patterns in Human Mammary Epithelial Cells and Breast Cancer," *Proceedings of the National Academiy of Sciences*, USA, 96, 9112-9217.

Phatak, A., and Reilly, P.M., and Penlidis, A. (1992), "The Geometry of 2-Block Partial Least Squares," *Communications in Statistics-Theory and Methods*, 21, 1517-1553.

Press, S.J. (1982), *Applied Multivariate Analysis: Using Bayesian and Frequentist Methods of Inference*. Robert E. Krieger Publishing Company Inc., Malabar, Florida, 2nd edition.

Rocke, D.M. and Durbin, B. (2000), "A Model for Measurement Error for Gene Expression Arrays," to appear in *Journal of Computational Biology*.

Ross et al. (2000), "Systematic Variation in Gene Expression Patterns in Human Cancer Cell Lines," *Nature Genetics*, 24, 227-235.

Scherf et al. (2000), "A Gene Expression Database for the Molecular Pharmacology of Cancer," *Nature Genetics*, 24, 236-244.

Stone, M., and Brooks, R. J. (1990), "Continuum Regression: Cross-validated Sequentially Constructed Prediction Embracing Ordinary Least Squares, Partial Least Squares, and Principal Components Regression" (with discussion), *Journal of the Royal Statistical Society*, Series B, 52, 237-269.

9

APPLYING CLASSIFICATION SEPARABILITY ANALYSIS TO MICROARRAY DATA

Zhen Zhang[1,2], Grier Page[1], and Hong Zhang[3]

[1] *Department of Biometry and Epidemiology, Medical University of South Carolina, Charleston, SC 29425.* : [2] *3Z Informatics, Mt. Pleasant, SC 29466.* : [3] *BIOwulf Technologies, Savannah, GA 31405.*

Abstract: We describe a novel approach to process genome-wide expression data from multiple arrays representing different classes of experiment conditions. We first derive a new unified maximum separability analysis (UMSA) procedure for constructing linear classifiers and demonstrate that the procedure unifies the classic linear discriminant analysis method and the optimal margin hyperplane method as used in support vector machines. We then present a stepwise backward algorithm using UMSA to compute significance scores for individual genes based on their collective contribution to the separation of different classes of arrays. Using the public data sets of the budding yeast saccharomyces cerevisiae, we demonstrate the effectiveness of the UMSA based algorithms in identifying genes with the most discriminatory power in separating arrays of cells under normal division cycles and those under heat shock.

Key words: Expression data processing, separability analysis, microarray data analysis, SVM, discriminant analysis.

INTRODUCTION

Two basic approaches have been commonly used in published papers for processing genome-wide expression data from microarray experiments. The first involves the use of algorithms based on the principal component analysis (PCA) method to reduce dimensionality of the expression data [Raychaudhuri et al, 2000, Alter et al, 2000]. The second uses a variation of cluster analysis algorithms to organize the expression data so that genes sharing similar expression profiles across multiple experiments and/or experiments sharing similar expression profiles across the entire set of genes are arranged close together for easy identification of expression variation patterns [Eisen et al, 1998].

PCA based methods are characterized by the ability to decouple and decorrelate multivariate data. When PCA is used for dimension reduction, it retains maximum information for *data representation*. It is an excellent tool to extract information from raw data to provide a global picture of the dynamics of gene expression.

However, in genome-wide expression analysis, it is quite often that the experiment conditions under which the individual array data are obtained have been selected or designed with a particular endpoint of interest in mind. The purpose of the analysis is to identify those genes whose change in expression levels can be related back to the known designations of experiment conditions. An example of such analysis would be the comparison of microarray expression data of specimens from a particular type of tumor cells against controls from normal tissues. For such data, it is quite possible that those genes with the largest expression variance might not be associated with the end point of interest in the analysis at all. In other words, the so called principal components might not necessarily be pointing to the directions along which the different classes of arrays are best separated.

Cluster analysis provides a holistic view of expression patterns across the entire set of observations. A similar drawback of this approach is, however, that the majority of genes in the data set, including those with strong expression variation patterns, might not be associated at all with the predefined end point of interest. Consequently, expression patterns that are truly associated with the end point would have to have a strong presence in terms of the number of genes of similar profiles so that they could be detected among the large number of non-contributing expression patterns.

Both PCA and cluster analysis are, using terminology from pattern recognition and machine learning, unsupervised methods. These methods do not use information from the designated labels of data points.

In this paper, we first derive a new unified maximum separability analysis (UMSA) procedure for constructing a multivariate linear classifier that optimally separates two classes of data according to a predefined separability measure. We then present a supervised algorithm using the UMSA procedure in a backward stepwise fashion to iteratively assign significance scores to individual genes that rank them according to their *collective* contributions to the separation of different classes of arrays. We argue that for many genome-wide expression data processing problems, especially those in clinical or biomedical research, a supervised algorithm that incorporates information from data labels may provide results that are more relevant to the predefined research goals.

UNIFIED MAXIMUM SEPARABILITY ANALYSIS

The construction of a linear classifier for separating two classes of data in an n-dimensional space may be equivalently viewed as to find an $R^n \to R$ linear projection, represented by a normalized $n \times 1$ unit projection vector d, such that a predefined separability measure between the two classes of data is maximized. Furthermore, the entries in d represent the relative contributions of the n original variables in a collective fashion towards the maximum separation of the classes.

Fisher's linear discriminant analysis (LDA) has been widely applied in pattern classification for its simplicity and robustness [Fisher, 1923]. When two classes of data are normally distributed and share the same covariance structure, LDA offers the optimal separation between the two classes. Under LDA, the separability measure to be maximized is the distance between the two class means divided by the pooled variance, all in the projected one-dimensional space:

$$s = \left|d^T(M_2 - M_1)\right| / d^T \Sigma d \qquad [1]$$

where Σ is the pooled covariance matrix, and M_i's are the class means, all in the original n-dimensional space. It is easy to show that the maximum value is $s_{max} = \|\Sigma^{-1}(M_2 - M_1)\|$ with the corresponding unit projection vector:

$$d = \Sigma^{-1}(M_2 - M_1) / \left\|\Sigma^{-1}(M_2 - M_1)\right\| \qquad [2]$$

It is difficult, however, to directly apply LDA to microarray data for separation of arrays from different experiment conditions. For most data sets, the number of genes >> the number of arrays available. The estimation of covariance matrix Σ becomes a severely ill-posed problem [Friedman, 1989].

LDA is only one of several possible ways to construct linear classifiers. Another widely used type of linear classifier is based on the concept of optimal margin hyperplane which has been an essential component of the support vector machine (SVM) classifiers [Vapnik, 1998]. Let D denote the set of all $R^n \to R$ linear projections of norm 1. For two linearly separable classes of data ω_1 and ω_2, the optimal margin hyperplane classifier is determined by finding a unit project vector d such that the maximal margin between the classes is assumed:

$$s_{\max} = \max_{d \in D} \left\{ \min_{x_i \in \omega_1, x_j \in \omega_2} \left| d^T (x_i - x_j) \right| \right\} \qquad [3]$$

With some modification (soft margin [Vapnik, 1998]), optimal margin hyperplane classifiers can be also used for linearly non-separable data.

It should be noted that an optimal margin hyperplane classifier is determined solely by data points close to the separation boundaries between the two classes and does not explicitly use information about distributions of data, which makes it easily applicable for cases such as genome expression data where the dimensional of space is much greater than the number of data points. For the same reason, however, such methods tend to be very sensitive to mislabelled data. For microarray data for biomedical research, labelling errors are likely to occur due to misclassified or contaminated specimens.

In this paper, as the foundation of an algorithm for expression data processing, we derive a new procedure called unified maximum separability analysis (UMSA) that allows for the incorporation of partial data distribution information into the determination of a modified optimal margin hyperplane classifier. We further show that with adjustable parameters, the UMSA procedure produces classifiers that encompass both the LDA classifiers and the optimal margin hyperplane classifiers as two of its extreme cases.

Considering the expression data e as m column vectors $x_1, x_2, \ldots, x_m$ and their corresponding class membership as $c_1, c_2, \ldots, c_m \in \{-1, +1\}$, the expression in [3] for an optimal margin hyperplane classifier for linearly separable data may be reformulated as a solution to the following optimization problem:

$$\begin{aligned} &\text{Maximize } s = (b_2 - b_1) \\ &\text{Subject to} \\ &\quad d \cdot x_i \le b_1, \text{if } c_i = -1 \\ &\quad d \cdot x_i \ge b_2, \text{if } c_i = 1 \\ &\quad \|d\| = 1, b_1 \le b_2, \end{aligned} \qquad [4]$$

where b_1 and b_2 are two scalars in the one dimensional projection space. Define $b = -(b_1 + b_2)/(b_2 - b_1)$ and $\upsilon = 2d/(b_2 - b_1)$, and [4] becomes:

$$\begin{aligned}&\text{Minimize } \|\upsilon\| \\ &\text{Subject to} \\ &\quad \upsilon \cdot x_i + b \le -1, \text{if } c_i = -1 \\ &\quad \upsilon \cdot x_i + b \ge 1, \text{if } c_i = 1\end{aligned} \qquad [5]$$

which can be equivalently expressed as:

$$\begin{aligned}&\text{Minimize } \frac{1}{2}\upsilon \cdot \upsilon \\ &\text{Subject to} \\ &\quad c_i(\upsilon \cdot x_i + b) \ge 1,\ i = 1,2,\ldots,m\end{aligned} \qquad [6]$$

After solving [6], we have $s_{\max} = 2/\|\upsilon\|$, and the $n \times 1$ unit projection vector $d = \upsilon/\|\upsilon\|$ [Vapnik, 1998].

To accommodate situations where the two classes are not completely separable and to deal with noisy data, outliers, or mislabelled data, we relax the constraints by introducing non-negative variables $\xi_1, \xi_2, \ldots, \xi_m$ that represent errors in the constraints. These errors are penalized in the objective function:

$$\begin{aligned}&\text{Minimize } \frac{1}{2}\upsilon \cdot \upsilon + \sum_{i=1}^{m} p_i \xi_i \\ &\text{Subject to} \\ &\quad c_i(\upsilon \cdot x_i + b) \ge 1 - \xi_i,\ i = 1,2,\ldots,m\end{aligned} \qquad [7]$$

where coefficients p_i are positive constraints representing the relative importance of individual data points. It should be noted that if p_i are set to be a constant for all data points, equation [7] has the same form and solution as the so called optimal soft margin hyperplane linear classifier described in the SVM literature [Vapnik, 1998].

Let δ_i be the shortest distance between a data point x_i and the line that goes through the two class means, M_1 and M_2. We set p_i to be inversely related to δ_i through a positive decreasing function $\phi(\cdot)$:

$$p_i = K\phi(\delta_i), \qquad [8]$$

where K is a constant. A typical choice for the function $\phi(\cdot)$ would be

$$\phi(x) = e^{-x^2/\sigma^2} \qquad [9]$$

which is also the one used in this paper. The quadratic program in [7] has a unique solution. By recasting it based on Kuhn-Tucker theorem, a number of proposed efficient algorithms may be employed [Vapnik, 1998].

The parameters K in [8] and σ in [9] alter the hyperplane resulting from UMSA. The constant K plays the same role as in a regular SVM to dictate the maximum amount of influence a data point may exert over the minimization process and hence controls the "softness" of the optimal margin. The new parameter σ, on the other hand, regulates the amount of information on data distribution being incorporated into the determination of υ. A large σ relative to the spread of data will essentially turn [9] into a constant. The result will be a regular optimal soft margin hyperplane classifier. On the other hand, a very small σ causes any data points further away from the line between the class means to quickly lose their significance in the objective function and become less contributory towards the determination of the hyperplane. The corresponding classifier will be close to a regular LDA function with assumed identity covariance matrix. (For non-identity covariance cases, the computation of distance δ needs to be normalized by the pooled covariance matrix. We choose not to do so in our algorithms for expression data processing due to difficulties in estimating the covariance matrix when $n >> m$). In this sense, the value $s = 2/\|\upsilon\|$ from solving [7], itself now a function of K and σ, may be viewed as a relative measure of linear separability between two classes of data. The selection of the two parameters K and σ should be based on information about the distribution of the data and the noisiness of the data and their labels.

In a typical parametrical statistical classifier, data points contribute equally to the estimation of distribution parameters such as Σ and M_i's in the case of LDA, independent of their locations relative to the class boundaries. On the other hand, using empirical risk minimization, classifiers such as the optimal margin hyperplane, rely solely on data points that are close to the class boundaries and essentially ignore the overall distribution of data classes. These two approaches, under the assumption of independently and identically distributed data, offer good performance when the training data sets are sufficiently large. For many current biological expression data analysis problems, however, the sample sizes are often extremely small. For such problems, the two approaches are not very efficient in utilizing all information from all data points.

The expressions in [7]-[9] may be seen as merely a minor change to the SVM formulation. It is, however, an important attempt to unify two historically and traditionally different approaches for pattern classification. It allows for the efficient use of information from data points near boundaries

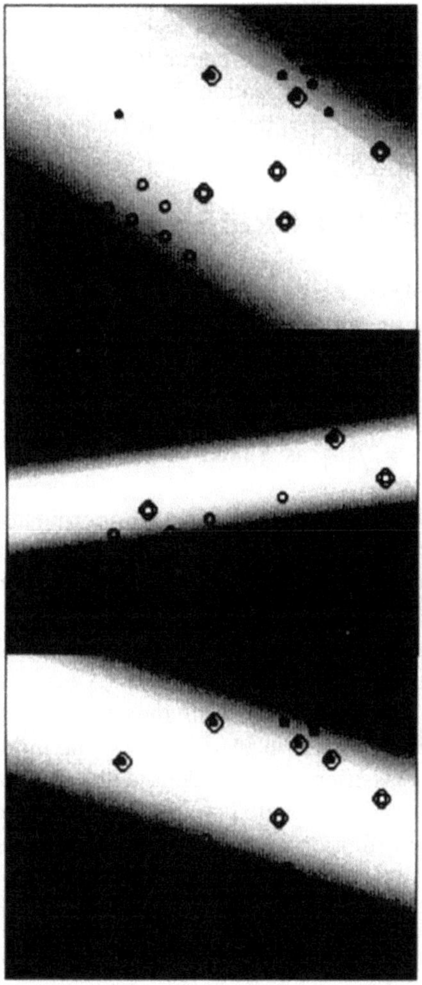

Figure 1. Effect of UMSA parameter σ. In all sub-figures, solid and hollow points represent two classes of made-up samples. Changing shades indicate location and orientation of classification boundaries. Points with outer circles are "support vectors" from optimal margin computation for a given set of UMSA parameters K and σ. Top: UMSA using a extremely small σ, the classification boundary is similar to that from LDA; Middle: UMSA using a large σ (relative to the spread of data), the classification boundary is a pure optimal margin classifier determined by boundary point only; Bottom: UMSA using a small σ (yet comparable to the spread of data), the classification boundary is determined by both boundary points and overall data distribution.

while at the same time applying distribution-related information to improve performance against noise and labelling errors in data. In Figure 1, the effect of a varying parameter σ is illustrated. It shows that by allowing the incorporation of distribution information (the direction through the two class means), the resultant classification boundary can be made much less sensitive to a few out-of-place points in the training data.

AN ITERATIVE ALGORITHM FOR GENE SELECTION

Based on UMSA, a gene selection algorithm can be constructed by repeatedly eliminating least significant genes. A significance measure of genes in terms of their contributions in the classification is required for the algorithm.

It is easy to prove that for a fixed selection of UMSA parameters K and σ, the quantity $s = 2/\|v\|$ as a relative linear separability measure has the following property:

Let $A_n = \{1,2,\cdots,n\}$ be the index set of coordinates of an n dimensional vector space and A_k a k element subset of A_n. For a given two-class data set D_n in an n dimensional space with index set A_n and its linear projection D_k in a subspace formed by the coordinates specified in A_k, we have $s_k \le s_n$, where s_k and s_n are separability measures computed from applying UMSA on D_k and D_n, respectively.

The value s_k provides an overall measure of separability for D_k. To facilitate the gene selection, we need to assess the individual contributions of the genes in the classifier. Let d_k denote the unit projection vector from applying UMSA on D_k with components d_k^j, $j \in A_k$, along its jth dimension. The contribution of dimension j in separating the two classes in D_k is

$$h_k^j = s_k \cdot \left| d_k^j \right| \qquad [10]$$

In general, h_k^j is not necessarily a monotonic function of k. The boundary data points that contribute most to the determination of the UMSA classifier in D_{k-1} may be partially different from those supporting the UMSA classifier in D_k. To appropriately rank the contributions of individual dimensions to the separation of data in D_n, we use a backward stepwise procedure. At each step a single dimension r is recursively eliminated from A_k according to its computed significance score. Define

$$w_k^j = \max\{h_k^j, h_{k+1}^j, \ldots, h_n^j\} \qquad [11]$$

for all dimension $j \in A_k$ and

$$w_k^* = \min_{j \in A_k}\{w_k^j\} \quad [12]$$

Let $r \in A_k$ be a dimension such that $w_k^r = w_k^*$ and we define w_k^r as the backward stepwise significance score of dimension r. The subset A_{k-1} is then obtained by eliminating dimension r from A_k. w_k^* is a monotonic function of k, since by definition, for a subsequently eliminated dimension r', we have:

$$w_{k-1}^* = w_{k-1}^{r'} \geq w_k^{r'} \geq w_k^r = w_k^* \quad [13]$$

For a given expression data set in an n-genes$\times$ m-array matrix e, the following algorithm implements the backward stepwise procedure in [11]-[12]. The returned vector w contains the computed significance scores of the n genes representing the relative importance of the genes in separating the two predefined classes of array experiments:

inputs:

UMSA parameters K and σ;

data $e = \{e_{ji} | j = 1,2,\ldots,n; i = 1,2,\ldots,m\}$; and

class labels $C = (c_1, c_2, \ldots, c_m)$, $c_i \in \{-1,+1\}$.

initialization:

$G_k \leftarrow G_n = \{g_j = (e_{j1}, e_{j2}, \ldots, e_{jm})^T, j = 1,2,\ldots,n\}$;

score vector $w = (w^1, w^2, \ldots, w^n)^T \leftarrow (0,0,\ldots,0)^T$.

while $|G_k| > 1$

1 forming $X = (x_1, x_2, \ldots, x_m) \leftarrow (g_1, g_2, \ldots, g_k)^T$.

2 applying UMSA(K,σ) on X and C;

$s_k \leftarrow 2/\|v\|$ and $d_k \leftarrow v/\|v\|$.

3 for all $g_j \in G_k$, if $s_k |d_k^j| > w^j$, $w^j \leftarrow s_k |d_k^j|$.

4 $G_{k-1} \leftarrow G_k - \{g_r\}$, where r is determined from

$$w^r = \min_{g_j \in G_k}\{w^j\}.$$

return w.

RESULTS

We applied the *iterative algorithm for gene selection* with UMSA parameters $K = 5.0$ and $\sigma = 2.5$ to rank and sort genes in the budding yeast saccharomyces cerevisiae data set [Eisen et al, 1998] with the purpose of identifying genes that respond to heat shock. The significance scores in this case are computed according to the contribution of individual genes in a collective fashion to the separation of array experiments of yeast during normal division cycles and those under heat shock at different time points.

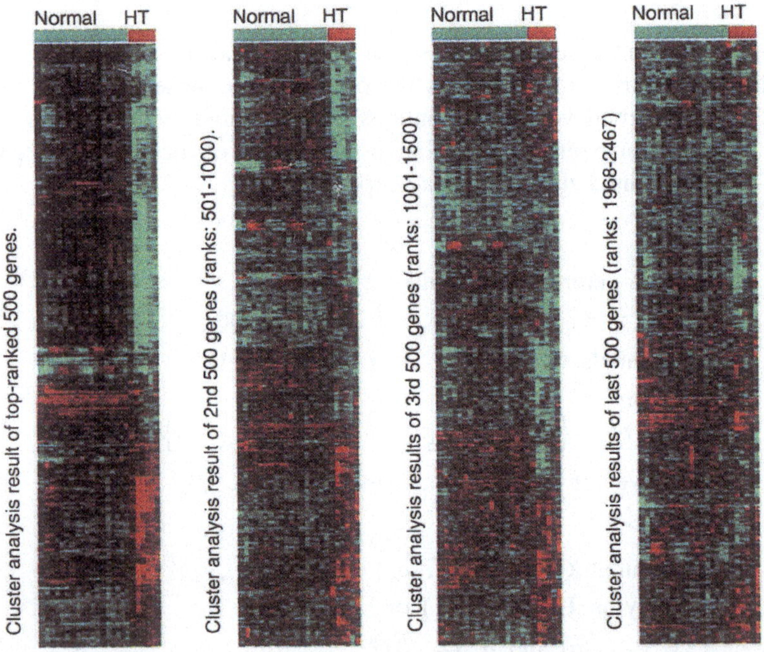

Figure 2. Clustering results of sorted genes, in batches of 500 genes each, selected according to the computed significance score for their contribution to the separation of experiments of yeast under normal division cycles and under heat shock stress (HT) at different time points.

First we sort the entire set of genes according to their computed significance score. We then use the Gene Cluster and Treeview programs from Stanford University [Eisen et al, 1998] on groups of the top-ranked 500 genes, the second and third ranked 500 genes, and finally the least significant 500 genes. Figure 2 shows that the top-ranked 500 genes, after clustering, demonstrate expression patterns over a large number of genes that are strongly associated with the heat shock stress. Such association decreases rapidly and progressively towards the end of the sorted genes, with the last 500 genes showing almost completely random patterns after clustering. This shows that the scoring algorithm effectively "filters" out the genes with expression variation patterns that are closely associated with the difference in array experiments.

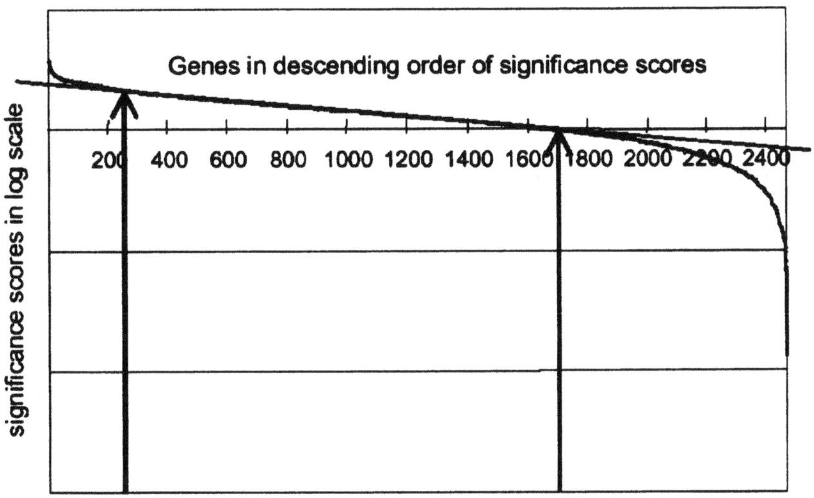

Figure 3. Log-scale significance score in descending order of 2467 genes in the yeast data set. Change point of descending rate may be used to objectively determine number of genes to select.

In Figure 3, the log-scaled significance scores of 2467 genes in the yeast data set are plotted in descending order. It is easy to see that the first 250 plus genes are the most significant ones in separating the two classes of array experiments. Afterwards their significance decreases exponentially (the fitted straight line). The last 700 plus are essentially non-contributory towards the data separation. Such a plot provides an objective tool to determine the proper number of genes to select for further analysis. With this observation, we select 250 top-ranked genes to perform a more detailed cluster analysis. The clusters of genes from our analysis seem to have responses to heat shock that are biologically plausible. Among them, the down regulated clusters are composed primarily of ribosomal proteins, as expected since protein synthesis is strongly down regulated in heat shock conditions, and genes involved in RNA metabolism. The clusters of strongly up regulated genes appear to have two classes of genes: the class of heat shock proteins, chaperonins, and other genes involved in cellular and protein stability, and the class of genes involved in metabolism and the uptake of metabolites. In response to heat shock, the cells are not making many new proteins, and at the same time are trying to protect the existing proteins and to expand the list of possible energy sources.

CONCLUSIONS

Currently, the effective use of analytical methods for microarray data processing is often hindered by the extremely high dimensionality (number of genes) and the small sample sizes. The efficient use of all available information becomes a critical issue. In this paper, we presented a supervised algorithm to assign significance scores to individual genes in a microarray data set based on their contribution to the separation of specimens of different classes. The underlying UMSA procedure improves information utilization by incorporating estimated partial data distribution information into a structural risk minimization procedure. Using the publicly available yeast data set, we demonstrated the usefulness of the algorithm in filtering out genes that are most relevant to the phenotypic differences between classes of specimens.

REFERENCES

Alter, O., Brown, P.O., & Botstein D. (2000) *Proc. Natl. Acad. Sci. USA* **97**, 10101-10106.

Eisen, M.B., Spellman, P.T., Brown, P.O., & Botstein D. (1998) Cluster analysis and display of genome-wide expression patterns, *Proc. Natl. Acad. Sci. USA* **95**, 14863–14868.

Fisher, A. (1923) *The Mathematical Theory of Probabilities, Vol. 1*, (Macmillan, New York).

Friedman, J.H. (1989) Regularized discriminant analysis, *J. Am. Stat. Assoc.*

Raychaudhuri, S., Stuart, J.M., & Altman, R.B. (2000) *Pacific Symposium on Biocomputing* **5**:452-463.

Vapnik, V.N. (1998) *Statistical Learning Theory*, (John Wiley & Sons, New York), pp.401-441.

10

HOW MANY GENES ARE NEEDED FOR A DISCRIMINANT MICROARRAY DATA ANALYSIS

Wentian Li and Yaning Yang
Laboratory of Statistical Genetics, Box 192, The Rockefeller University, 1230 York Avenue, New York, NY 10021, USA.

Abstract: The analysis of the leukemia data from Whitehead/MIT group is a discriminant analysis (also called a supervised learning). Among thousands of genes whose expression levels are measured, not all are needed for discriminant analysis: a gene may either not contribute to the separation of two types of tissues/cancers, or it may be redundant because it is highly correlated with other genes. There are two theoretical frameworks in which variable selection (or gene selection in our case) can be addressed. The first is model selection, and the second is model averaging. We have carried out model selection using Akaike information criterion and Bayesian information criterion with logistic regression (discrimination, prediction, or classification) to determine the number of genes that provide the best model. These model selection criteria set upper limits of 22~25 and 12~13 genes for this data set with 38 samples, and the best model consists of only one (no.4847, zyxin) or two genes. We have also carried out model averaging over the best single-gene logistic predictors using three different weights: maximized likelihood, prediction rate on training set, and equal weight. We have observed that the performance of most of these weighted predictors on the testing set is gradually reduced as more genes are included, but a clear cutoff that separates good and bad prediction performance is not found.

Key words: model/variable selection, Akaike/Bayesian information criterion, maximum likelihood, logistic regression, discriminant analysis

INTRODUCTION

There are two types of microarray experiments. In the first type, samples are not categorized, but gene expression levels are followed with time. The goal is to find genes whose expression levels vary together. In the second type, samples are labelled as either normal or diseased tissues. The goal is to find genes whose expression levels can distinguish different labels. Using terminology from machine learning, the data analysis of the first type of experiment is unsupervised, whereas that of the second type is supervised (see, e.g., [Haghighi et al., 1999] for this particular application, and [Ripley, 1996] for general discussions). The prototype of the first type of analysis is cluster analysis, whereas that of the second type is discriminant analysis.

We focus on the data analysis of the second type, using the leukemia data from Whitehead/MIT's group [Golub et al., 1999]. The question we address is: How many expressed genes are needed for a discriminant analysis? In this data set, 7129 gene expressions are measured on 38 sample points. In the original analysis [Golub et al., 1999], 50 out of more than 7000 gene expressions are used for discriminant analysis. We ask whether 50 genes are still too large a number, considering that the sample size is only 38. This seemingly simple question may receive two answers from two different perspectives: model selection (see, e.g., [Burnham and Anderson, 1998]) and model averaging (see, e.g., [Geisser, 1993]). We will address the two separately below.

MODEL SELECTION

The goal of model selection is to pick the model among many possible models that achieves the best balance between data-fitting and model complexity. A perfect data-fitting performance by a complicated model with many parameters can well be an example of overfitting. On the other hand, a simplistic model with few parameters that fits the data poorly is an example of underfitting. There are at least two proposals for achieving the balance between data-fitting and model complexity: Akaike information criterion (AIC) [Akaike, 1974; Parzen et al., 1998; Burnham and Anderson, 1998] and Bayesian information criterion (BIC) [Schwarz, 1976; Raftery, 1995]. These two quantities are defined as

$$AIC=-2\log(\hat{L})+2K \quad \text{and} \quad BIC=-2\log(\hat{L})+\log(N)K \qquad [1]$$

where $\hat{L}$ is the maximized likelihood [Edwards, 1972], K the number of free parameters in the model, and N the sample size. High-order terms of $O(1/N)$ (for AIC) and $O(1) + O(1/\sqrt{N}) + O(1/N)$ (for BIC) are ignored here for simplicity. The model with the lowest AIC or BIC is considered to be the preferred model.

Table 1. Logistic regression results (sample size N=38, log(N)=3.637568) K: number of free parameters in the logistic regression (LR) (number of genes included plus 1); ΔAIC (ΔBIC) is the AIC (BIC) value relative to that of the best model) single-gene LR using g4847, zyxin); p_{train} (p_{test}) is the success (prediction) rate on the training (testing) set; "top 37" is the LR using 37 best genes by their single-variable LR performance. Notes (a) Since this model/predictor fits the data perfectly, L should be 1, and $-2\log(\hat{L})$ should be 0. In a real optimization procedure, the actual value may depend on the number of iterative steps. (b) This LR is selected by a stepwise variable selection (by either AIC or BIC) from the starting group of top 22 genes (top22 → AIC/BIC → 2). (c) Similar to (b), but the starting group of genes in the LR containing the top 10 genes (top10 → AIC/BIC → 2). (d) This null model 1 uses 11/38 ≈ 0.29 as the probability for AML for any sample data point. Since 0.29 < 0.5, any sample will be predicted as ALL type, which is correct 27 times in the training set, and 20 times in the testing set. The $-2\log(\hat{L})$ is equal to $-2\log[(27/38)^{27}(11/38)^{11}]$. (e) The $-2\log(\hat{L})$, p_{train}, p_{test} for the null model 0 (random guess model) is expected to be:$-2\log(0.5^N)= 2N\log(2)$, 0.5, and 0.5. A more complete gene description and the GenBank accession number of most genes listed here can be found in the Appendix.

Model	K	$-2\ln\hat{L}$	AIC	ΔAIC	BIC	ΔBIC	P_{train}	P_{test}
1 g4847(zyxin)	2	~0[a]	4	0	7.28	0	38/38	31/34
2 g1882(CST3 cystatin C)	2	6.97	10.97	6.97	14.25	6.97	36/38	32/34
3 g3320(c4 synthase)	2	10.91	14.91	10.91	18.19	10.92	35/38	27/34
4 g5039(leptin receptor)	2	11.36	15.36	11.36	18.63	11.36	36/38	22/34
5 g6218(ELA2 elastatse 2)	2	12.46	15.46	12.46	18.73	11.46	34/38	22/34
6 g2020(FAH...)	2	12.10	16.10	12.10	19.38	12.10	36/38	25/34
7 g1834 (CD33 antigen)	2	12.23	16.23	12.23	19.50	12.23	35/38	31/34
8 g760 (cystatin A)	2	13.10	17.10	13.10	20.38	13.10	35/38	32/34
9 g1745 (LYN v-yes-1...)	2	13.15	17.15	13.15	20.43	13.15	33/38	28/34
10 g5772 (c-myb)	2	14.72	18.72	14.72	22.00	14.72	35/38	27/34
100 g2833 (AF1q)	2	27.22	31.22	27.22	34.49	27.22	30/38	28/34
200 g3312 (p.k. ATR)	2	30.84	34.84	30.84	38.12	30.84	29/38	21/34
g1834+g2267 [b]	3	0.004	6.00	2.00	10.92	3.64	38/38	22/34
g5039+g5772 [c]	3	0.008	6.01	2.01	10.92	3.65	38/38	26/34
Top2(g4847+g1882)	3	0.029	6.03	2.03	10.94	3.67	38/38	32/34
Top5	6	0.011	12.01	8.01	21.84	14.56	38/38	24/34
Top10	11	0.002	22.00	18.00	40.02	32.74	38/38	31/34
Top22	23	0.001	46.00	42.00	83.67	76.39	38/38	27/34
Top37	38	0.001	76.00	72.00	138.2	130.9	38/38	21/34
null 1 (proportion)	1	45.73[d]	47.73	43.73	49.73	42.09	27/38[d]	20/34[d]
null 0 (random guess)	0	52.68[e]	52.68	48.68	52.68	45.40	19/38[e]	17/34[e]

Selecting genes relevant to a discriminant microarray data analysis may become an issue of model selection. But we have to be clear in what context this is the case. Model selection adjusts the number of parameters in the model, not the number of variables. Nevertheless, if we plan to combine variables (additively or any other functional form), each variable has one or perhaps more coefficients. Removing a variable removes the corresponding coefficients. In this context, variable selection is a special case of model selection. We illustrate this by the logistic regression/discrimination for our data set:

$$\Pr(AML) = \frac{1}{1 + \exp(-a_0 - \sum_{j \in \text{top genes}} a_j x_j)} \qquad [2]$$

where x_j is the (log, normalized) gene expression level of gene *j* among the top performing genes, and AML (acute myeloid leukemia) is one type of leukemia (ALL, the acute lymphoblastic leukemia, is another type). Using a linear combination of all 7129 genes in logistic discrimination requires 7130 parameters, whereas using one gene requires only 2 parameters. AIC/BIC defined in [1] will then compare multiple-gene models with single-gene models, and determine which scenario is better.

Our results are summarized in Table 1, where we list the type of model (which variables are additively combined in the logistic regression), number of parameters in the model, -2log of the maximum likelihood, AIC/BIC (absolute and relative), success rate in training set, and prediction rate in the testing set. The most striking result from this data set is that many genes are strong predictors for the leukemia class, consistent with the observation in [Golub et al., 1999]. This observation is confirmed by the following facts in Table 1:

•Logistic regressions using the top 2, 5, 10, 22,and 37 genes all fit the data almost perfectly (as measured by the -2 log($\hat{L}$) value: the closer is $\hat{L}$ to 1, the nearer to perfect fit) (note that the model is saturated when the number of variables used in the logistic regression is 37, since the number of parameters is then equal to the number of sample points).

•Stepwise variable selection leads to only two genes (note that stepwise variable selection fails to find the best model, a single-gene model, because it is a local minimization/maximization procedure).

•The No. 10 best-performing gene is only slightly worse than the No.1 performing gene: 35 vs. 38 correct predictions on the training set (though the No. 100 and No. 200 best-performing genes predict only 30 and 29 correctly on the training set).

This situation of strong prediction or easy classification is in contrast with the epidemiology and pedigree data used in human complex (non-

Mendelian) diseases, where strong predictors are rather rare [Li et al., 2000; Li and Nyholt, 2001].

The last two rows in Table 1 are predictors/classifiers that do not use any gene expression levels. These are called null models. The first null model (1) uses the proportion of AML in all training samples as Prob(AML) (11/38 in our data set) as the guessing probability for AML. The second null model (0) uses half-half probabilities. It is interesting to note that to beat both null models, the number of genes in logistic regression can not exceed certain upper limits: for null model 1, since we require AIC/BIC to be smaller (where p is the number of genes used in logistic regression, and $-2 \log(\hat{L})$ is assumed to be zero for the best-case scenario):

$$0 + 2(p+1) \approx AIC < AIC(null\ 1) = 47.728$$
$$0+3.637568\ (p+1) \approx BIC < BIC(null\ 1) = 49.365, \qquad [3]$$

it sets $p < 22.86 \approx 22$, and $p < 12.57 \approx 12$, respectively. Similarly, to beat null model 0, we require:

$$0+2(p+1) \approx AIC < AIC(null\ 0) = 52.679$$
$$0+3.637568\ (p+1) \approx BIC < BIC(null\ 0) = 52.679 \qquad [4]$$

which are $p < 25.34 \approx 25$, and $p < 13.48 \approx 13$, respectively. Any logistic regression using more than these number of genes (22 ~ 25, 12 ~ 13), will have a larger value of AIC or BIC than the two null models. Since by the model selection criterion, only models with smaller values of AIC/BIC are selected, we should not include more than these number of genes in our logistic regression. Note that these specific upper limits are directly related to our particular sample size (i.e., 38). For larger sample sizes, the upper limit for the number of genes will be larger.

MODEL AVERAGING

In the model selection framework, we can not use a logistic regression with too many genes because it may not improve the data-fitting performance enough to compensate for the increase of model complexity. This conclusion is correct when one model (e.g. a logistic regression that uses one gene) is compared to other alternative models (e.g. a logistic regression that uses, say, 10 genes). Nevertheless, it is possible to average/combine many different models each involving one gene. The restriction on model complexity during the model selection process does not apply to model averaging. Model averaging has also been discussed under similar names such as "committee machines", "boosting weak predictors", "mixture of experts", "stacked generalization" (see, e.g., [Ripley, 1996]).

Without guidance from the model selection framework on the number of models (number of genes) to be included, we have tried several empirical approaches. We first examine whether there is a gap in data-fitting performance among top genes. Genes that do not fit the data should not be considered in model averaging. For this purpose, Figure 1 shows the -2 log($\hat{L}$) for the top 1000 genes on the training set, as a function of the (log) rank. A linear fitting of -2log($\hat{L}$) on log(rank) seems to fit the points well, and it is hard to see "better-than-average" genes except the first gene. This is not surprising since g4847 discriminates the 38 training sample points perfectly. Note that the linear trend in Figure 1 is equivalent to a power-law function in likelihood vs. rank plot ($\hat{L} \sim 1/r^{2.56}$). Such power-law is similar to the power-law rank-frequency plots observed in many social and natural data, also known as Zipf's law [Zipf, 1949; Li, 1997-2001]. In any case, there is no discernible gap in Figure 1 that separates relevant and irrelevant genes (more discussions on Zipf's plots similar to Figure 1, but obtained from other microarray data can be found in [Li, 2001]).

We then check how the prediction rate on the training set correlates with that on the testing set. Figure 2 shows the error rates on both training (x-axis) and testing sets (y-axis) for the top 500 performing genes. The left plot in Figure 2 shows the mean square error, and the right plot shows the prediction errors. The left plot contains both information on the success rate of prediction and that of confidence of prediction, whereas the right plot contains only information on the success rate of prediction. Points along the diagonal line in Figure 2 exhibit similar error rates in the training and testing set, and thus are reasonable predictors. On the other hand, points well above the diagonal line indicate overtraining. The most reasonable predictor based on both the training and testing set is g1882 (on the other hand, the best predictor based on the training set only is g4847).

Finally we examine the model averaging performance with various numbers of models included, each being a single-gene logistic regression:

$$\Pr(AML) = \sum_{j \in \text{top genes}} w_j \left(\frac{1}{1 + \exp(-a_j - b_j x_j)} \right) \qquad [5]$$

Although the terms to be summed in [5] can be any model, here we only examine the simplest case where the model to be averaged is the single-gene logistic regression model.

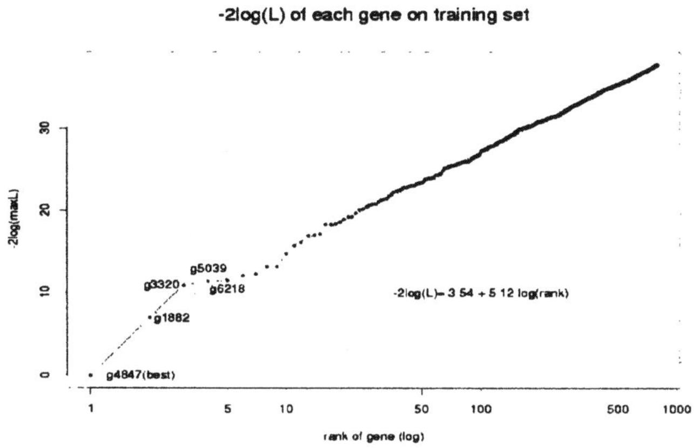

Figure 1. Data-fitting performance as measured by -2log of the maximized likelihood (the smaller, the better fit) for the top 1000 genes. The x-axis is the rank of the gene by the maximum likelihood of its single-gene logistic regression. A linear regression line of -2log( ) vs log (rank) is also shown, with the slope equal to 5.12.

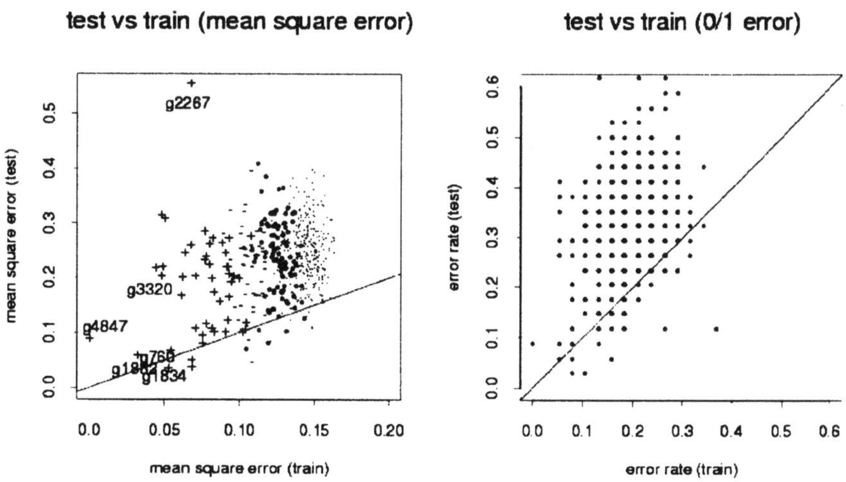

Figure 2. Error rates on the testing set (y-axis) are compared to those on the training set (x-axis) for the top 500 genes (based on the performance on the training set). Each point represents one single-gene logistic regression predictor. The left plot uses the mean squared error ("+" for the first top 100 genes, "-" for the rank 51-100 genes, "*" for the rank 101-200 genes, and "." For the rank 201-500 genes, and the right plot uses the prediction error.

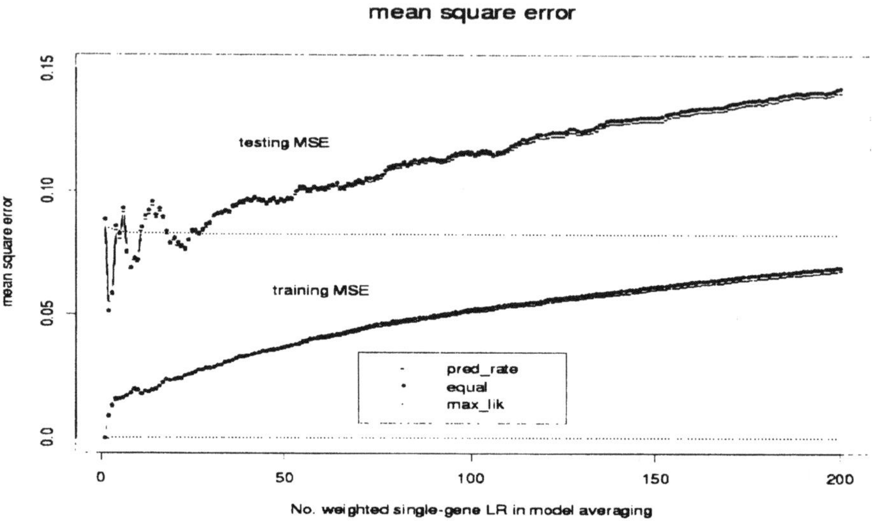

Figure 3. Model averaging performance with three different weighting schemes: (1) weight being proportional to the prediction rate on the training set; (2) equal weight; and (3) weight being proportional to the maximum likelihood obtained on the training set. The mean square error is plotted against the number of models being averaged. Each model is a single-gene logistic regression. The mean square error on both the training set (bottom) and the testing set (top) are shown.

We have chosen three weighting schemes: the first is proportional to the prediction rate on the training set ($w_j \propto p_{j,train}$), the second is the equal weight ($w_j \propto 1$), and the last one is proportional to the maximum likelihood as obtained from the training set ($w_j \propto \hat{L}_j$). Figures 3-4 show the behaviour of all three weighting schemes on both the training and the testing sets, using either the mean square error or the prediction error, up to 200 genes.

The maximum likelihood weight is equivalent to the Akaike weight ($w_j \propto \exp(-AIC_j/2)$) [Parzen et al., 1998] and Bayesian weight ($w_j \propto \exp(-BIC_j/2)$) [Raftery, 1995], since all models being averaged in [5] have the same number of parameters and same sample size (assuming high-order terms are ignored). Using Bayesian weight in model averaging is essentially to derive the posterior predictive distribution [Gelman et al., 1995]: "posterior" because data in a training set is used, and "predictive" because unknown new data in the testing set is to be predicted.

For our data set, this weighting scheme is nevertheless uninteresting: the mean square error only decreases slightly with the number of models used, while the error in prediction rate is unchanged. The reason for this is very simple - the best model (the single-gene logistic regression using the g4847) discriminates the training set perfectly. Its likelihood is much higher than any other models. As a result, the weight of other models is negligible, and the model averaging essentially remains as one model. The equal weight can be potentially incorrect since models that do not fit the data should not contribute as equally as models that fit the data better. In our data set, however, the equal weight scheme is actually similar to the weighting scheme that uses the prediction rate, because many genes (up to 200) exhibit similar prediction rates on the training set.

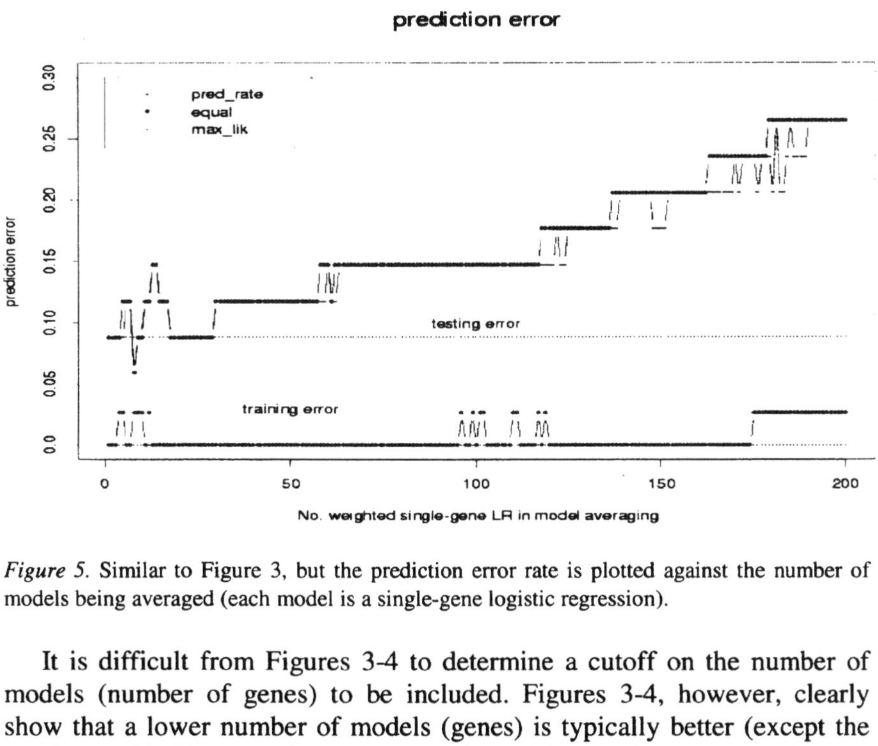

Figure 5. Similar to Figure 3, but the prediction error rate is plotted against the number of models being averaged (each model is a single-gene logistic regression).

It is difficult from Figures 3-4 to determine a cutoff on the number of models (number of genes) to be included. Figures 3-4, however, clearly show that a lower number of models (genes) is typically better (except the maximum likelihood weight, which is insensitive to the number of genes included). Figure 3, in particular, indicates that it is possible to perform better (in term of mean square error) on the testing set than using just one model if the number of models is less than around 25. Of course, this result is obtained from the specific testing set we have at hand. A more conclusive result may require a combining of the training and testing set, or requires more sample points than are currently available.

Due to the weight in a model averaging, the apparent number of terms (models, genes) included does not reflect on the true number of genes involved. For this reason, we may introduce a quantity called "effective number of genes (terms, models)". The leading term in a model averaging contributes a number of 1 to this quantity, but the contribution from the term j is equal to w_j/w_1. For example, with the maximum likelihood weight, the relative weights of the first ten terms are 1, 0.031, 0.0043, 0.0034, 0.0032, 0.0024, 0.0022, 0.0014, 0.0014, and 0.0006. The effective number of terms when the top ten genes are included is 1.05, a far less number than 10.

The discrimination used in [Golub et al., 1999] is a model averaging instead of a model selection. The number of genes used is 50. To quote from [Golub et al., 1999], "the number was well within the total number of genes strongly correlated with the class distinction, seemed likely to be large enough to be robust against noise, and was small enough to be readily applied in a clinical setting." Our Figures 3-4 show that although the prediction rate on the training set stays close to 100% even as the number of models (genes) is increased, the prediction rate on the testing set decreases with more genes. The only exception is the maximum likelihood weight, where the prediction rate is almost unchanged due to the dominance of the best gene. It seems that we should not increase the number of models (genes) in model averaging arbitrarily.

SUMMARY

It is important to emphasize what we have done/concluded and we have not. We have addressed the number of genes to be included in a predictor *when these genes are combined in a logistic regression model or any other model.* The answer to this question is found from the perspective of model selection. If gene expressions from different genes are not combined, but examined individually, how many genes should be claimed to be relevant to the discrimination of the two labels is a different issue, and usually involves a subjective decision. For example, in a t-test, the hypothesis that the expression of a particular gene is different in two types of samples is tested. Whether this hypothesis is accepted, or more accurately, whether the null hypothesis is rejected, depends on the significance level. Different choices of the significance level lead to various number of genes claimed to be "relevant".

In a model averaging framework, the question on the number of genes to be included in the discrimination is somehow ill phrased. Because of the concept of the weight, inclusion of one gene is not equal to an inclusion of another gene with less effect. There is no limit on the number of genes that could be included in a model averaging because adding expression values from unimportant genes does not greatly affect the nature of the predictor. In this context, the more natural question is '*what is the value of the effective number of genes?'*. It is very much possible that even if the expression of all 7129 genes is included in the predictor by a model averaging, the effective number of genes can be still as small as 2.

Another topic we did not address is the correlation between genes and the related concept of gene clusters. In this paper, the variable being selected is the individual gene, the model being averaged is logistic regression using an individual gene, etc. Nevertheless, all these analyses can be easily extended to situations where not only genes, but also clusters of genes, are targets of selection or averaging. In [Hastie et al., 2001], for example, the variable to be selected includes both the original genes and clusters at various levels obtained from a hierarchical cluster analysis.

We also did not address the possibility that the incorrect predictions are due to the sample outlier [Li and Wong, 2001] or sample heterogeneity. Table 1 shows that the best fitting model on the training set is the single-gene logistic regression using g4847 (zyxin). With this predictor, there are three incorrect predictions on the testing set: samples 66(AML), 67(T-cell ALL) and 68(B-cell ALL). The best predictor on the testing set is the single-gene logistic regression using g1882 (CST3 cystatin C), or that using g760 (cystatin A). With the g1882 predictor, there are only two incorrect predictions, on samples 61(AML) and 67(T-cell ALL); with g760, samples 61 and 66 (both AML) are incorrectly predicted. We notice that sample 67 is the only T-cell ALL in the testing set, so it is not impossible that it behaves different from the B-cell ALL's (see [Grant et al., 2001]).

In conclusion, due to the small sample size and the presence of strong predictors, we believe that the number of genes used in a discriminant analysis in this data set can be much smaller than 50 as stated in [Golub et al., 1999]. Although we can not give a definite answer as to the exact number of genes to be used, one proposal is to use only one or two genes, and other exploratory data analyses indicate an upper limit of 10-20 genes for this data set. A more extensive study of the same issue using other microarray data sets can be found elsewhere (Li et al., in preparation).

ACKNOWLEDGEMENTS

W.Li's work was supported by NIH grant K01HG00024 and Y. Yang's work was supported by the grant MH44292.

APPENDIX

Table 2. Gene names and GenBankAccession numbers for the genes in Table 1

Gene #	description	GenBank A#
4847	Zyxin	X95735_at
1882	CST3 Cystatin C (amyloid angiopathy and cerebral hemorrhage)	M27891_at
3320	Leukotriene C4 synthase (LTC4S) gene	U50136_rna1_at
5039	LEPR Leptin receptor	Y12670_at
6218	ELA2 Elastatse 2, neutrophil	M27783_s_at
2020	FAH Fumarylacetoacetate	M55150_at
1834	CD33 antigen (differentiation antigen)	M23197_at
760	Cystatin A	D88422_at
1745	LYN V-yes-1 Yamaguchi sarcoma viral related oncogene homolog	M16038_at
5772	C-myb gene extracted from Human, complete primary CDS, and five complete alternatively spliced CDS	U22376_cds2_s_at
2267	CDC25A Cell division cycle 25A	81933_at

REFERENCES

Akaike, H (1974), "A new look at the statistical model identification", IEEE Transactions on Automatic Control, 19:716-723.

Burnham, KP, Anderson, DR (1998), *Model Selection and Inference* (Springer).

Edwards, AWF (1972), *Likelihood* (John Hopkins Univ Press).

Geisser, S (1993), *Predictive Inference: An Introduction* (Chapman & Hall).

Gelman, A, Carlin, JB, Stern, HS, Rubin, DB (1995), *Bayesian Data Analysis* (Chapman & Hall).

Golub, TR, Slonim, DK, Tamayo, P, Huard, C, Gaasenbeek, M, Mesirov, JP, Coller, H, Loh, ML, Downing, JR, Caligiuri, MA, Bloomfield, CD, Lander, ES (1999),"Molecular classification of cancer: class discovery and class prediction by gene expression monitoring", Science, 286:531-537.

Grant GR, Manduchi E, Stoeckert Jr. CJ (2001), "Using non-parametric methods in the context of multiple testing to identify differentially expressed genes", Chapter 3, this volume

Haghighi, F, Banerjee, P, Li, W (1999), "Application of artificial neural networks in whole-genome analysis of complex diseases" (meeting abstract), Cold Spring Harbor Meeting on Genome Sequencing & Biology, page 75.

Hastie, T, Tibshirani, R, Botstein, D, Brown, P (2001), "Supervised harvesting of expression trees", Genome Biology, 2(1):research0003.1-0003.12.

Li C, Wong WH (2001), "Model-based analysis of oligonucleotide arrays: expression index computation and outlier detection", Proceedings of National Academy of Sciences, 98:31-36.

Li, W (1997-2001), A page on Zipf's law (URL: linkage.rockefeller.edu/wli/zipf/).

Li, W (2001), "Zipf's law in importance of genes for cancer classification using microarray data", submitted.

Li, W, Nyholt, D (2001), "Marker selection by Akaike information criterion and Bayesian information criterion", Genetic Epidemiology, in press.

Li, W, Sherriff, A, Liu, X (2000), "Assessing risk factors of complex diseases by Akaike information criterion and Bayesian information criterion" (meeting abstract), American Journal of Human Genetics, 67 (supp 2), page 222.

Parzen, E, Tanabe, K, Kitagawa, G (1998), eds. *Selected Papers of Hirotugu Akaike* (Springer).

Raftery, AE (1995), "Bayesian model selection in social research", in Sociological Methodology, ed. PV Marsden (Blackwells),pages 185-195.

Ripley, BD (1996), Pattern Recognition and Neural Networks (Cambridge University Press).

Schwarz, G (1976),"Estimating the dimension of a model", Annals of Statistics, 6:461-464.

Zipf, GF (1949), *Human Behavior and the Principle of Least Effect* (Addison-Wesley).

11

COMPARING SYMBOLIC AND SUBSYMBOLIC MACHINE LEARNING APPROACHES TO CLASSIFICATION OF CANCER AND GENE IDENTIFICATION

Werner Dubitzky, Martin Granzow, Daniel Berrar
German Cancer Research Center,
Intelligent Bioinformatics Systems Group,
Heidelberg, Germany

Abstract: Microarray experiments provide the scientific community with huge amounts of data. Without appropriate methodologies and tools significant information and knowledge hidden in these data may not be discovered. Therefore, there is a need for methods capable of handling and exploring large data sets. The field of data mining and machine learning provides a wealth of methodologies and tools for analyzing large data sets. Generally, such methods can be categorized into symbolic and subsymbolic methods. In the context of the acute myeloid leukemia (AML) and acute lymphoblastic leukemia (ALL) data sets provided by CAMDA 2000, we investigated and compared two representatives of classical machine learning approaches: *decision trees* (symbolic) and *artificial neural networks* (subsymbolic). The experiments indicated that the symbolic approach yields more accurate results than the neural network method.

Key words: Microarray analysis, gene expression profiling, decision tree, artificial neural network

INTRODUCTION

Classification of human tumors into distinguishable entities is preferentially based on clinical, pathohistological, enzyme-based histochemical, immunohistochemical, and in some cases cytogenetic data. Such an approach to classification often provides classes containing tumors that show similarities but differ strongly with respect to some important aspects, such as clinical course, treatment response, or survival. Data obtained from high-throughput experiments (e.g., cDNA microarrays for profiling gene expression in tissues) have the potential to address this problem. However, microarray experiments generate huge amounts of data that is frequently characterized by many more descriptors (independent variables) than observation entities (cases). This raises the question as to which methods could be reliably used to analyze such data.

Based on microarray gene expression data, we investigated the performance of two popular machine learning [Mitchell, 1997] techniques in the context of molecular classification of cancer and identification of potentially relevant genes. The techniques in question are *decision trees* and *artificial neural networks* (see Chapter 1 for a more general discussion of these techniques). Decision trees belong to the class of so-called symbolic methods. The attractiveness of decision trees is largely due to their ability to express the learned models as symbolic rules that can readily be understood by humans. Neural network approaches, on the other hand, represent their learned knowledge as patterns of connectivity that exist among the nodes of the network. This type of knowledge representation is sometimes called subsymbolic, and is not readily intelligible by humans. In their standard implementation both methods do not readily lend themselves to the determination of single-event probabilities of their predictions. However, this does not necessarily mean that these methods cannot be statistically validated.

Rather than devising special-purpose algorithms, our intention was to use standard, off-the-shelf software tools for analyzing the data provided for the CAMDA competition [CAMDA, 2000]. As a basis for our comparative study we chose two of the most popular algorithms currently available in proprietary machine learning and data mining software: the decision tree algorithm C5.0 and the well-known backpropagation algorithm for neural networks [Witten & Frank, 2000; Berry & Linoff, 1997; Dayhoff, 1996]. For both algorithms we used the proprietary implementation realized in the data mining tool Clementine from SPSS [SPSS]. Besides using standard tools, our motivation was to contribute to the understanding of the underlying techniques, their benefits and their limitations.

DESCRIPTION OF PROBLEM

A recent study [Golub et al., 1999] analyzed gene expression data from a training set of 38 (27 ALL, 11 AML) and a test set of 34 (20 ALL, 14 AML) leukemia patients for class discovery (cluster analysis) and prediction (discriminant analysis). Out of 7,070 genes, they found 1,400 genes to be highly expressed in ALL or AML. From those they picked an arbitrary total of 50 genes with the highest association with disease type and used them for prediction. The aim of our analysis is to investigate

- the relative classification performance of a standard symbolic (decision tree C5.0) and a standard subsymbolic (backpropagation artificial neural network) machine learning method; and
- the value of these methods with respect to their capability to identify the most relevant genes.

DESCRIPTION OF EXPERIMENTS

The general approach of our analysis was to directly use all expression data (except the control data) without further processing.

Firstly, we aimed at comparing the performance of the neural network and decision tree methods on the basis of (a) a 5-fold cross-validation procedure applied to *all* 72 cases plus the original training/test set (henceforth referred to by cross-validation index cv0), and (b) a validation only on the basis of the provided training/test data split. Secondly, we aimed at identifying relevant genes based on the best-performing model. And thirdly, we aimed at finding a way of usefully combining neural networks and decision trees.

As both the backpropagation neural network algorithm and the C5.0 decision tree algorithm allow some control of the learning behavior through parameter settings, we decided after some exploratory screening to run three sets of experiments with different parameter configurations for either approach. So we did three times 6-fold cross-validation runs with different decision tree configurations and three times 6-fold cross-validation runs with different neural networks configurations.

Cross-Validation

We have randomly subsampled the entire set of 72 cases into five training sets (n=15) and five test sets (n=57). Henceforth, these cross-validation samples are referred to by the cross-validation indices: cv1, cv2, ..., cv5. In addition to these five training/test set pairs, we used the original training set (n=38) and test set (n=34) as the sixth cross-validation pair; this is referenced by the cross-validation index cv0. The distribution of the tumors for these six cross-validation training/test set pairs is shown in Table 1.

Table 1. Distribution of tumors over the cross-validation data sets.

Tumor	cv 0	cv 1	cv 2	cv 3	cv 4	cv 5
	Training Sets					
AML	11	17	19	21	20	21
ALL	27	40	38	36	37	36
total	38	57	57	57	57	57
	Test Sets					
AML	14	8	6	4	5	4
ALL	20	7	9	11	10	11
total	34	15	15	15	15	15

To establish the performance of the six models and parameter configurations (3 times decision tree, 3 times neural network), we trained each model/parameter configuration six times (on each of the six training sets) and applied the resulting classifier to the corresponding test set. Then we computed the individual performances for each test set and the average performance, using absolute accuracy and *lift* measures, of each model/parameter configuration over the six test runs. See tabular presentation of these results in Tables 2 and 3.

To determine the relative performances between the decision tree and neural network approach, we picked the best-performing (according to average performance over six test runs) of the three decision tree classifiers and the best-performing of the three neural network classifiers and established through *t*-statistics if the performance of the two is statistically significant. Please refer to the section on Statistical Validation of Best-Performing Models for a detailed description of the methods used.

Table 2. Cross-validation performances.

	ANN Test Set Performance			**Decision Tree Test Set Performance**		
cv / size	**Acc. Th: 90%**	**Acc. Th: 85%**	**Acc. Th: 80%**	**Standard**	**10-fold boo.**	**20-fold boo.**
cv 0; n=34	79.41*	79.41	70.59	91.17	91.17	91.18
cv 1; n=15	86.67	80.00	46.67	66.66	86.66	86.66
cv 2; n=15	86.67	86.67	93.33	80.00	100.00	100.00
cv 3; n=15	60.00	93.33	60.00	93.33	100.00	100.00
cv 4; n=15	66.67	73.33	66.67	86.66	86.67	86.66
cv 5; n=15	93.33	93.33*	93.33*	86.66	86.66	93.33
Avg	**78.79**	**<u>84.35</u>**	**71.77**	**84.08**	**91.86**	**<u>92.97</u>**

*Legend: Acc. Th. = training accuracy threshold; boo. = boosting; * = pre-set threshold range could not be achieved.*

Table 3. Summary of results.

	Neural Network						
Test Set(s)	**Accuracy %**			**Lift**			**Parameters**
	AML	ALL	TOT	AML	ALL	TOT	
Best CAMDA; cv = 0	**50.00**	**100.0**	**79.41**	**2.42**	**1.25**	**1.84**	**a = 87.5% (no >90%)**
Best All (cv = 0 .. n)	67.08	94.63	84.35	2.61	1.55	1.98	85% < a < 90%
Best Single Test Set	75.00	100.0	93.33	3.74	1.25	2.50	cv=5; a = 91.7%
All 3 Configurations	54.74	92.76	78.30	2.04	1.52	1.66	
	Decision Tree						
Test Set(s)	AML	ALL	TOT	AML	ALL	TOT	
Best CAMDA; cv = 0	**92.85**	**90.00**	**91.18**	**2.10**	**1.61**	**1.85**	**Std., 10fb, 20fb**
Best All (cv = 0 .. n)	92.56	92.62	92.97	2.64	1.54	2.09	20-fold boosting
Best Single Test Set	100.0	100.0	100.0	3.74	1.36	2.56	cv=2, cv=3 10/20fb
All 3 Configurations	90.94	88.28	89.64	2.44	1.36	2.56	

Legend: 10fb = 10-fold boosting; 20fb = 20-fold boosting

Parameter Configurations

The selection of model parameter configurations was largely determined by the implementation of the two approaches we chose (the Clementine data mining tool version 5.0.1 [SPSS]) and some initial exploratory experimentation.

C5.0 is the successor of C4.5 [Quinlan & Quinlan, 1997]. It is a complex and sophisticated decision tree algorithm with many parameter options. After some testing we chose to use the following three settings:

- *Standard.* In this mode the algorithm does not attempt to perform any boosting.
- *10-fold boosting.* Here, the algorithm attempts 10-fold boosting. Notice, depending of the actual performance of individual runs, there may be less than 10 models generated.
- *20-fold boosting.* Same as above, but with a fold of 20.

For all other setting we used the default values: *Gain Ratio* = ON, *Pruning Severity* = 70%, and *Minimum Record Per Branch* = 2. We did not specify any non-default costs for misclassification.

The neural network algorithm in Clementine supports various algorithms and parameter options. Critical for us was the option *dynamic* = ON, as this mode dynamically searches and evaluates different network topologies. After some experimentation we decided that option *Stop On* = *Accuracy* is most critical. If this option is selected one must specify a percentage value that determines a prediction accuracy threshold (for the training set). The learning process is terminated once this threshold is reached. To avoid overfitting, we chose the following training accuracy thresholds: 80%, 85%, and 90%. For all other settings we used defaults. Similar to the boosting approach, that does not guarantee *n* models when *n*-fold boosting is specified, the training accuracy may linger just below a threshold (e.g., 80%) and then jump to 98%. Thus, for a given training set, it may not be possible to obtain a final training accuracy within a preset accuracy *interval* (e.g., [80%,85%]).

This leads to three runs of 6-fold (cv0, ..., cv5) cross-validation analysis using the C5.0 decision tree model with the settings *Standard*, *10-fold boosting*, and *20-fold boosting*, and to three runs of 6-fold (cv0, ..., cv5) cross-validation with the neural network backpropagation algorithm using the training accuracy threshold settings 80%, 85%, and 90% respectively. See also Tables 2 and 3.

The Lift Measure

To glimpse into performance beyond simple accuracy or error measures, we used a measure frequently employed in machine learning called *lift*. The lift measures the strength of an effect against some expected "baseline". This measure takes into account the frequency of a class in the data set (test set), how often the class is predicted by the classifier, and how many *correct* predictions are made for the class.

Given the set of class labels, $C = \{c_1, ..., c_n\}$ and the set of cases, $S = \{x_1, ..., x_m\}$, let $act(x_j)$ denote the *actual* class (label) of case x_j and $prd(x_j)$ the class (label) *predicted* for x_j by a classifier. Then the *lift* for a particular class c_i, $lift(c_i)$, is measured by the prior probability, $p(act(x_j)=c_i)$, of class c_i occurring in S, and the conditional probability, $p(act(x_j)=c_i \mid prd(x_j)=c_i)$ of class $act(x_j)=c_i$ given the prediction, $prd(x_j)=c_i$, as follows:

$$lift(c_i) = \frac{p(act(x_j) = c_i \mid prd(x_j) = c_i)}{p(act(x_j) = c_i)} \qquad [1]$$

For example, let there be 15 cases in the test set, S, and let 3 of them have the class label A. So $p(act(x_j)=\text{A}) = 3/15 = 1/5$. If a model predicts the class A for 6 out of these 15 cases, and gets 2 of them correct, then $p(act(x_j)=\text{A} \mid prd(x_j)=\text{A}) = 2/6 = 1/3$. And then $lift(\text{A}) = (1/3) / (1/5) = 5/3 = 1.66$. Notice, the lift of 1.66 is much better than the lift of 1.00 which would be achieved if the model predicted class A for *all* elements in the set S, that is $p(act(x_j)=\text{A} \mid prd(x_j)=\text{A}) = 3/15 = 1/5$. This would of course lead to an accuracy (for class A) of 100%, giving a false impression of the actual performance of the model.

Statistical Validation of Best-Performing Models

One of the main intentions of this study was to compare the performance of a standard decision tree and a standard neural net within the context of microarray data analysis. To do so, we first selected the best performing decision tree and neural configuration with respect to the performance on *all* test sets, cross-validation indices cv0 to cv5 (see Table 2). The neural network performed best by allowing it to train to an accuracy of more than 85% and less than 90%, and the best-performing decision tree model was that with 20-fold boosting. We computed the z-value according to equation [2], the *mean prediction error*, $merr(M;X)$, of the model, M, over all predictions in the common test set, $X = \{x_1, \ldots, x_n\}$, according to equation [3], the individual prediction error, $err(M;x_i)$, of the model, M, for an individual case, x_i, according to equation [4], the combined standard error, $se(A;X,B;X)$, of two models, A and B, on the common test set, X, according to equation [5], and the combined variance, $var(A;X,B;X)$, in accordance with equation [6]. Finally, the p-value was taken from a standard *t*-statistics table. Table 7 shows a summary of these statistics. This procedure is also described in [Weiss and Indurkhya, 1998].

$$z = \frac{|merr(A;X) - merr(B;X)|}{se(A;X,B;X)} \quad [2]$$

$$merr(M;X) = \frac{\sum_{i=1}^{n} err(M;x_i)}{n} \quad [3]$$

$$err(M;x_i) = \begin{cases} 0 \text{ if } prd(M;x_i) = act(x_i) \\ 1 \text{ if } prd(M;x_i) \neq act(x_i) \end{cases} \quad [4]$$

$$se(A;X,B;X) = \sqrt{\frac{var(A;X,B;X)}{n}} \quad [5]$$

$$\text{var}(A;X,B;X) = \frac{1}{n}\sum_{i=1}^{n}\left[\left(err(A;x_i) - err(B;x_i)\right) - \left(merr(A;X) - merr(B;X)\right)\right]^2 \quad [6]$$

In equations [2] to [6], n refers to the number of cases in the data set (test set X), and $act(x_i)$ denotes the *actual* class of case x_i and $prd(x_i)$ the class *predicted* for x_i by a classifier or model M (or A or B).

Identification of Genes

To identify which genes are most important, we first identified the best-performing decision tree and neural configuration with respect to the performance on the given CAMDA 2000 test set (n=34), cross-validation index cv0. For the decision tree approach *all* tree configurations yielded identical performance (31 out of 34 correctly classified), and for the neural network approach the configuration training accuracy threshold = 85% performed best (27 out of 34 correctly classified). In fact, within the given time limits for training, we could not get the network to achieve a training performance of more than 87.5% (see also Table 2 and Table 3.) The genes we identified are presented in the section on Results and Discussion.

RESULTS AND DISCUSSION

We first look at the individual results and performances of the two approaches and then compare and contrast them. Consult Tables 2 and 3 for this discussion.

Neural Network Results and Discussion

The best classification performance of the neural network method was obtained by interrupting the backpropagation learning process between 85% and 90% (average: 88.43%) predicted accuracy. In this case, the average classification accuracy over all 6 cross-validation runs was 84.35%. Training the net to a predicted accuracy, x, of $x > 90\%$ and $80\% < x < 85\%$ respectively resulted in lower actual prediction performances (78.79% in the former and 71.77% in the latter case).

Further analysis showed that although for each of the three neural net runs the ALL tumor was classified with a higher accuracy than the AML class: ALL avg. classification accuracy over all three runs: 92.76%, for AML: 54.74%. However, the lift measure for the AML class scored higher in each of the test runs: ALL avg. lift score over all three runs: 1.52, for AML: 2.04. This means that the model showed a significantly higher sensitivity/selectivity with regard to the AML class. See also Tables 2 and 3 for a detailed breakdown of these results.

Training times for each neural network model was limited to a maximum of 5 minutes (on a 2 x Pentium III PC), and the obtained network topologies ranged from the most complex (6 nodes in the first, and 4 in the second hidden layer) to the least complex (2 nodes in the first, and 2 in the second hidden layer). The network topologies for the best-performing neural network configuration (85% < predicted accuracy < 90%) were 7070-6-4-1, 7070-3-3-1, 7070-4-2-1, 7070-2-2-1, 7070-5-4-1, and 7070-5-3-1 respectively. Incidentally, the topology notation follows the following neuron layer pattern: ***input-hidden1-hidden2-output.***

Through sensitivity analysis we were able to identify which genes have the highest impact (relative importance) on the classification performance of the neural network. In Table 4 we provide the first 15 genes of a sensitivity analysis carried out on the *entire* data set of 72 cases, based on a 7070-3-2-1 topology and a learning accuracy threshold setting of 90%. The full list can be obtained from the authors.

Table 4. Identified genes through neural network sensitivity analysis.

Gene Description	Gene Accession #	Sensitivity
ELA2 Elastatse 2, neutrophil	M27783_s_at	0.00902
(clone S31i125) mRNA, 3' end of cds	U61734_s_at	0.00838
PPGB Protective protein for beta-galactosidase (galactosialidosis)	M22960_at	0.00827
Homeotic Protein C6, Class I	HG3921-HT4191_f_at	0.00823
GB DEF = Myotubularin related protein 3 (MTMR3) gene, partial cds	U58034_at	0.00804
Oncoprotein 18 (Op18) gene	M31303_rna1_at	0.00797
Catalase (EC 1.11.1.6) 5'flank and exon 1 mapping to chromosome 11, band p13 (and joined CDS)	X04085_rna1_at	0.00797
GB DEF = GPSAT=glycophorin SAT [human, peripheral bloods, mRNA Partial, 407 nt]	S77893_s_at	0.00793
OBF-1 mRNA for octamer binding factor 1	Z49194_at	0.00791
LGALS3 Lectin, galactoside-binding, soluble, 3 (galectin 3) (NOTE: redefinition of symbol)	M57710_at	0.00787
GB DEF = HH2A/c gene	Z83742_at	0.00784
TCF11 Transcription factor 11 (basic leucine zipper type)	X77366_at	0.00779
KIAA0022 gene	D14664_at	0.00776
S100A2 gene, exon 1, 2 and 3	Y07755_at	0.00774

Decision Tree Results and Discussion

The best classification performance of the C5.0 decision tree method was obtained on the basis of 20-fold boosting (combination of multiple different models). In this case the average classification accuracy over all 6 cross-validation runs was 92.98%. The result for 10-fold boosting was only marginally lower (91.87%). However, the non-boosting version of the decision tree only achieved an average classification accuracy of 84.09%. Interestingly, for the CAMDA 2000 training set (n=38) the boosting method was not able to derive multiple models, but repeated the known [Golub et al., 1999] result: Zyxin (accession code X95735_at) with an expression level of 938 as decision boundary. However, for many of the other cross-validation subsamples, boosting was able to identify multiple complementary models, thus indicating multiple genes and expression levels related to differentiating AML and ALL. The listing in Figure 1 shows a section of the trained decision trees for cv5 and 20-fold boosting. Indentations in the diagram indicate lower nesting levels, and the values in round brackets refer to *support* and *confidence* measures respectively.

```
Rule #1 - estimated accuracy 98.2% [boost 100.0%]:
    M23197_at =< 309 (34.0, 1.0) -> ALL
    M23197_at  > 309 (22.0, 0.955) -> AML
Rule #2 - estimated accuracy 96.4% [boost 100.0%]:
    M23197_at =< 390
        X71125_at =< 64 (38.6, 1.0) -> ALL
        X71125_at  > 64 (3.8, 0.605) -> AML
    M23197_at  > 390 (13.6, 1.0) -> AML
Rule #3 - estimated accuracy 94.6% [boost 100.0%]:
    M23197_at =< 390 (44.7, 0.957) -> ALL
    M23197_at  > 390 (11.3, 1.0) -> AML
Rule #4 - estimated accuracy 94.6% [boost 100.0%]:
    M31523_at =< 528 (26.2, 0.958) -> AML
    M31523_at  > 528 (29.8, 0.98) -> ALL
Rule #5 - estimated accuracy 92.9% [boost 100.0%]:
    L09209_s_at =< 911 (31.1, 1.0) -> ALL
    L09209_s_at  > 911 (24.9, 0.916) -> AML
...
```

Figure 1. Part of C5.0-generated decision tree for cv5, 20-fold boosting.

Further analysis showed that over all three C5.0 decision tree runs the AML class was classified with a higher accuracy than ALL (see Tables 2 and 3), with an average classification accuracy over all three runs: 90.94% for AML, and 88.28% for ALL. Moreover, the lift measure for the AML class scored significantly higher in each of the three test runs (ALL avg. lift score over all three runs: 1.50, for AML: 2.44). This means that the C5.0 decision tree model not only showed a significantly higher sensitivity/selectivity with regard to the AML class (when compared with ALL), but also a slightly higher precision. With regard to the ALL class both models showed comparable results regarding lift (sensitivity/selectivity) and precision (accuracy), but for AML the decision tree method clearly outperformed the neural net approach.

On an Intel-based PC (2 x Pentium III 600 MHz, 512 MB memory), training times of the C5.0 decision tree model construction ranged from 10-20 seconds for the non-boosting to 10-30 seconds for 10-fold boosting to approximately 100 seconds for 20-fold boosting.

Although in principle easy to implement, the identification of a full list of relevant genes with the decision tree implementation provided by Clementine turned out to be cumbersome. Furthermore, it was not possible to quantify the relative difference of the ranked genes. In Table 5 we provide a short list of genes (accession numbers) that were found to be most influential to the classification performance of the best decision tree model (20-fold boosting).

Table 5. Identified genes through decision tree analysis.

Gene Description	Gene Accession #
KIAA0181 gene, partial cds	D80003_at
CYSTATIN A	D88422_at
Glutamate Decarboxylase 1	HG2160-HT2230_at
Interferon gamma up-regulated i-5111 protein precursor	L07633_at
APLP2 Amyloid beta (A4) precursor-like protein 2	L09209_s_at
CD33 CD33 antigen (differentiation antigen)	M23197_at
CST3 Cystatin C (amyloid angiopathy and cerebral hemorrhage)	M27891_at
PTX3 Pentaxin-related gene, rapidly induced by IL-1 beta	M31166_at
TCF3 Transcription factor 3 (binding factors E12/E47)	M31523_at
GB DEF = Amelogenin (AMELX) gene, 3' end of cds	M55418_at
DF D component of complement (adipsin)	M84526_at
GLUTATHIONE S-TRANSFERASE, MICROSOMAL	U46499_at
Melanoma growth stimulatory activity (MGSA)	X54489_rna1_at
TST Thiosulfate sulfurtransferase (rhodanese)	X59434_at
Max gene extracted from H.sapiens max gene	X66867_cds1_at
Zyxin	X95735_at

Decision Tree vs. Neural Net: Results and Discussion

After running 2 x 3 x 6 cross-validation tests, we picked the best-performing neural network (85% < predicted accuracy < 90%) and decision tree model (20-fold boosting) and compared them using *t*-statistics. It is perhaps interesting to first inspect a detailed breakdown of misclassification of the two best-performing models in Table 6. The best-performing decision tree misclassified three cases, and the best neural net failed to correctly classify seven cases. The only case misclassified by both models is case# 66, which gives rise to the assumption that this case was not labeled correctly in the first place or that it belongs to a new subcategory.

Table 6. Breakdown of misclassifications on CAMDA test set (n=34).

prd(ANN85)	prd(DT 20fb)	Actual Tumor Label	Case#
ALL	AML	ALL	68
ALL	AML	ALL	67
ALL	AML	AML	52
ALL	AML	AML	54
ALL	AML	AML	57
ALL	AML	AML	60
ALL	AML	AML	65
ALL	AML	AML	62
ALL	ALL	AML	66

The overall performances of the two models are depicted in Figure 2 (in the diagram, the numbers on the X-axis correspond to the discussed cross-validation indices as follows: 1 = cv0, 2 = cv1, ..., 6 = cv5).

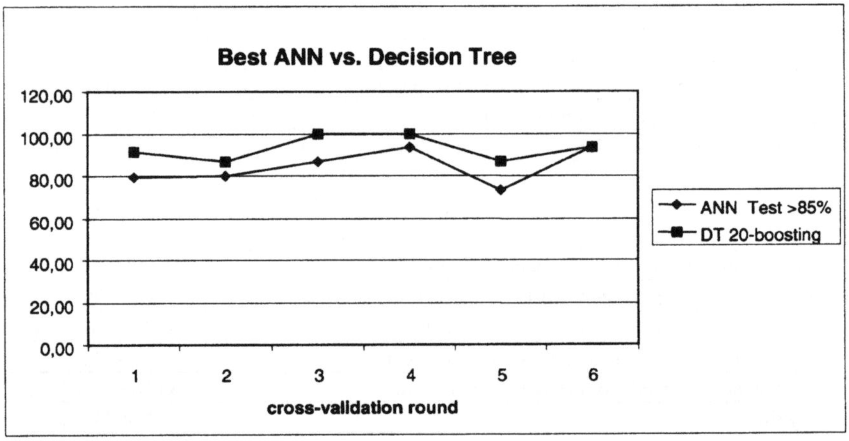

Figure 2. Overall classification performances of best-performing neural network and decision tree model (6-fold cross-validation).

It is easily seen from the diagram in Figure 2 that the best-performing decision tree model outperformed the best-performing neural network model (except in cv5 where both models fared equally). However, to decide if the performance differences are statistically significant is only possible after some statistical analysis. We carried out the analysis described in the section on Statistical Validation of Best-Performing Models [Weiss and Indurkhya, 1998]. Table 7 shows the results of this analysis.

Table 7. Statistics of best-performing models.

cv / size	merr ANN85	merr DT20	comb. SE	z-value	p-value
cv 0; n=34	0,21	0,09	0,08	3,64	< 0.01
cv 1; n=15	0,20	0,13	0,11	2,92	< 0.01
cv 2; n=15	0,13	0,00	0,09	1,52	n.s.
cv 3; n=15	0,07	0,00	0,06	1,04	n.s.
cv 4; n=15	0,27	0,13	0,13	3,11	< 0.01
cv 5; n=15	0,07	0,07	0,09	1,41	n.s.
avg	0,16	0,07	0,09	2,27	?

Legend: merr = mean error of model; ANN85 = neural net model with training accuracy threshold of 85%; DT20 = decision tree model with 20-fold boosting; comb. SE = combined standard error.

The results of the significance tests indicate that there are good reasons to believe that the performance difference of the two best-performing models is due to some systematic difference in their underlying learned mechanisms (rather than to chance variation). However, it would require further and more comprehensive tests to decide this with more confidence. The following is a list of observations and interpretations of the data analysis:

- both approaches can be used directly (no further preprocessing or discretization needed) with high-dimensional inputs (> 7000 genes) for molecular tumor classification and gene identification;
- the C5.0 decision tree classification model (a) shows higher precision and sensitivity levels, (b) provides an output format that is easy to interpret by humans (symbolic rules), and (c) is faster to train than the neural model; however, in the implementation we used it was cumbersome to generate a ranked list of important factors (genes) and quantify their relative importance;
- the neural network approach was treated unfairly in that it was not used in boosting (as this is not directly supported by the Clementine implementation); one could expect better performance (albeit with high computational costs) if the neural approach is used within a boosting framework;
- through sensitivity analysis the neural network approach was simpler and more precise with regard to ranking (plus relative performance) and identifying high-impact variables; however, this analysis carries high computational costs;

- Although significance analysis ranks all genes according to "impact" on classification, the analysis carries high computational cost when the number of genes is high ($n \sim 7000$);
- On the CAMDA training set (n=38) the decision tree was not able to come up with boosted (i.e., multiple) models as X95735_at (Zyxin) provided a perfect "split" that resulted in the following perfect-confidence rules: rule 1: ***if*** *X95735_at ≤ 938* ***then*** *ALL (support=27.0, confidence=1.0)*, and rule 2: ***if*** *X95735_at > 938* ***then*** *AML (support=11.0, confidence=1.0)*. To identify additional high-impact rules required to remove X95735_at and re-run the learning step on the remaining factors. This procedure eventually yielded the factors listed in *Table 5.*

With regard to the genes found by the two approaches we can make the following observations/interpretations. Two among the 10 genes identified by the decision tree encoded for (1) CD33 differentiation antigen that is tissue specific for monocytic/myeloid lineage cells (accession# M23197_at), and (2) transcription factor 3 (E2A) (accession# M31523_at). This is part of the described fusion transcript PBX1-E2A occurring in B-cell precursor ALL with translocation t(1;19)(q23;p13.3). Among the 50 most important genes identified by the neural network, several belong to:

- *ras-oncogene pathway*: Oncoprotein 18/Stathmin, MAPKAP kinase 3 (3pK), HCK (Hemopoietic cell kinase)) (e.g., accession# U09578_at);
- *Hox-gene family*: HoxA9 (an Oncogene), HoxB3 and the potential effector (Retinoic acid receptor beta (RARb)) (e.g., accession# U82759_at);
- *apoptosis associated genes*: LTBP1, a member of the *TGF-ß signal pathway,* and Interleukin 16, a cytokine activating the SAPK (a member of the MAPK family) signaling pathway (e.g., accession# M34057_at);
- *cell-specific genes:* OBF-1/POU2AF1 that is involved in a form of B-cell leukemia, and LGALS3 that is widely distributed in various tissues with the notable exception of B- and T-lymphocytes (e.g., accession# M57710_at); and
- *potential tumor suppressor gene:* hSNF2ßb/BRG-1 (chromatin remodeling) (e.g., accession# D26156_s_at).

SUMMARY AND CONCLUSION

Based on the CAMDA 2000 competition ALL/ALL data [CAMDA, 2000; Golub et al., 1999], we compared two classical machine learning approaches with regard to their potential of identifying genes and their performance in terms of discriminant analysis. The experiments indicated that the symbolic (decision tree) approach yields more accurate results than the neural network method in terms of classification performance. However, we found the sensitivity analysis of the neural approach useful to obtain a ranked list of "influential" genes. Clearly, much more analysis experiments, statistical and biological validation is necessary to establish a comprehensive set of properties on machine learning approaches to microarray data analysis.

REFERENCES

Berry, M.J.A. and Linoff, G., "Data Mining Techniques: For Marketing, Sales, and Customer Support", John Wiley & Sons, 1997.

CAMDA 2000 Web site, http://bioinformatics.duke.edu/camda/camda00/.

Dayhoff, J.E., "Neural Network Architectures: An Introduction", Thomson Computer Press, 1996.

Golub T.R., Slonim D.K., Tamayo P., Huard C., Gaasenbeek M., Mesirov J.P., Coller H., Loh M.L., Downing J.R., Caligiuri M.A., Bloomfield C.D., Lander E.S., "Molecular classification of cancer: class discovery and class prediction by gene expression monitoring", *Science* 286(5439) pp531-537, 1999.

Mitchell, T.M., "Machine Learning", McGraw-Hill, Singapore, 1997.

Quinlan, J.R. and Quinlan, R., "C4.5: Programs for Machine Learning", Morgan Kaufmann Series in Machine Learning, 1997.

SPSS: http://www.spss.com/datamine/, and Clementine User Group: http://www.spss.com/clementine/clug/ .

Weiss, S.M. and Indurkhya, N., "Predictive Data Mining: A Practical Guide", Morgan Kaufmann, Pub., San Francisco, CA, 1998.

Witten, I.H. and Frank, E., "Data Mining: Practical Machine Learning Tools and Techniques with Java Implementations", Morgan Kaufmann Pub., 2000.

12

APPLYING MACHINE LEARNING TECHNIQUES TO ANALYSIS OF GENE EXPRESSION DATA: CANCER DIAGNOSIS

Kyu-Baek Hwang, Dong-Yeon Cho, Sang-Wook Park, Sung-Dong Kim, and Byoung-Tak Zhang {kbhwang, dycho, swpark, sdkim, btzhang}@bi.snu.ac.kr
Biointelligence Laboratory, School of Computer Science and Engineering, Seoul National University, Seoul 151-742, Korea

Abstract. Classification of patient samples is a crucial aspect of cancer diagnosis. DNA hybridization arrays simultaneously measure the expression levels of thousands of genes and it has been suggested that gene expression may provide the additional information needed to improve cancer classification and diagnosis. This paper presents methods for analyzing gene expression data to classify cancer types. Machine learning techniques, such as Bayesian networks, neural trees, and radial basis function (RBF) networks, are used for the analysis of the CAMDA Data Set 2. These techniques have their own properties including the ability of finding important genes for cancer classification, revealing relationships among genes, and classifying cancer. This paper reports on comparative evaluation of the experimental results of these methods.

Key words: Bayesian networks, neural trees, radial basis function (RBF) networks, gene expression data analysis

INTRODUCTION

The treatment of cancer depends on its type. Accordingly, classification of patient samples is a crucial aspect of cancer diagnosis and treatment. It has been suggested that gene expression may provide the additional information needed to improve cancer classification and diagnosis. Recent technological advances in monitoring gene expression have led to a dramatic increase in gene expression data. Expression chips manufactured using technologies derived from computer-chip production can now measure the expression of thousands of genes simultaneously. A method for performing cancer classification using gene expression data is presented in [Golub et al., 1999] and [Slonim et al., 2000].

In this paper, we present methods for performing classification of patient samples by gene expression data analysis using machine learning techniques. Data-driven prediction must be able to extract essential features from individual examples and to discard unwanted information. Machine learning techniques are known to be excellent at discarding and compacting redundant information. We apply Bayesian networks, neural trees, and radial basis function (RBF) networks to cancer classification based on gene expression data. Among these methods, Bayesian networks are a kind of probabilistic graphical model. Neural trees and RBF networks are extensions of conventional neural networks and useful for regression.

The CAMDA Data Set 2 involves classifying acute leukemias. Acute leukemias can broadly be divided into two classes, acute myeloid leukemia (AML) and acute lymphoblastic leukemia (ALL). The dataset consists of 38 training samples and 34 test samples. Each sample comes from a patient and has 7,129 attributes that correspond to human gene expression levels. From the machine learning point of view, the cancer classification problem is formulated as follows. First, a model is constructed from training samples using a machine learning algorithm. Then, performance of the model is measured by test sample classification.

Our experimental results show that all training samples are classified correctly. The classification accuracy for test samples is also very high though it varies depending on the applied machine learning techniques. In addition to the classification accuracy, compacting redundant gene expression information and the discovery of relationships among genes are important and useful. In this sense, Bayesian networks are very interesting since they have the ability to disclose the probabilistic relationships among genes. Neural trees are also interesting since they can discover important genes for classification automatically, during the evolutionary process of model building.

This paper is organized as follows. The second section describes cancer classification with Bayesian networks and its experimental results. In the third and fourth sections, descriptions of cancer classification with neural trees and radial basis function (RBF) networks and their experimental results are given. Then, we compare and evaluate the experimental results of three machine learning techniques on the cancer classification problem. Conclusion and future work are presented in the last section.

MINING WITH BAYESIAN NETWORKS

Bayesian Networks

The Bayesian network represents the joint probability distribution for a set of random variables efficiently based on the concept of conditional independence. A Bayesian network assumes a form of directed acyclic graph (DAG). Figure 1 shows an example Bayesian network structure. Each node in the graph corresponds to a random variable and each edge represents the probabilistic dependency between variables. A Bayesian network which consists of n nodes (variables), $\mathbf{X} = \{X_1, \ldots, X_n\}$, represents the joint probability distribution as follows:

$$P(\mathbf{X}) = \prod_{i=1}^{n} P(X_i \mid \mathbf{Pa}_i), \qquad [1]$$

where $\mathbf{Pa}_i$ is the set of parents of X_i in the network structure. $P(X_i \mid \mathbf{Pa}_i)$ is the local probability distribution related to the node X_i. The global structure of a Bayesian network encodes the conditional independence relationships among all variables and is called to be the qualitative part. Local probability distributions for all nodes constitute the quantitative part of the Bayesian network.

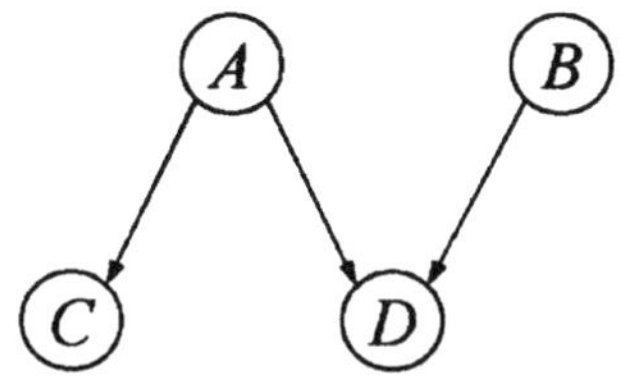

Figure 1. The directed acyclic graph (DAG) structure of an example Bayesian network that consists of four variables (nodes) A, B, C, and D. The joint probability distribution for these variables is represented as $P(A, B, C, D) = P(A)P(B|A)P(C|A, B)P(D|A, B, C)$ by the general chain rule. The network structure encodes the conditional independencies, such as $(A \perp B)$, $(C \perp B | A)$, and $(C \perp D | A, B)$ ($(X \perp Y | Z)$ denotes that X and Y are independent given Z). Using these conditional independencies, the joint probability distribution can be represented more compactly as $P(A, B, C, D) = P(A)P(B)P(C|A)P(D|A, B)$.

Because a Bayesian network encodes the joint probability distribution for a set of variables, the conditional probability of any interesting variable given observations of some of the other variables can be inferred efficiently. Therefore, once the Bayesian network whose nodes represent gene expression levels and the cancer class label is constructed from the gene expression data, the probability of the cancer class label given some gene expression levels for a new sample can be inferred. This is the Bayesian network classifier. [Jensen, 1996] and [Pearl, 1988] present efficient inference algorithms exploiting the structure of Bayesian networks. In addition to the classification, the edges in the Bayesian network structure denote the possibilities of causal relationships between variables.

Learning Bayesian networks from data is generally NP-hard [Chickering, 1996]. So, heuristic search algorithms, such as greedy search, greedy search with restart, and simulated annealing are used [Friedman and Goldszmidt, 1999].

Applying Bayesian Networks to the Cancer Classification Problem

In the analysis of CAMDA Data Set 2, each gene is regarded as a random variable. Thus, each gene corresponds to a node in a Bayesian network. There is an additional node for the leukemia class label (ALL or AML). The procedure of constructing a Bayesian network classifier for acute leukemias is as follows.

A node X_i in the Bayesian network has its local probability distribution $P(X_i \mid \mathbf{Pa}_i)$ and the local probability distribution model should be determined. In general, the unrestricted multinomial distribution model and the linear regression model are used for the local probability distribution of Bayesian networks. In this paper, the unrestricted multinomial distribution model is used. This model is very expressive. Also, learning algorithms and inference algorithms for the Bayesian network that has the unrestricted multinomial distribution, are established well. Dirichlet distribution is used for the parameters of the unrestricted multinomial distribution [Heckerman, 1996]. Gene expression levels are numerical values and must be discretized for the unrestricted multinomial distribution. In our experiments, all gene expression levels are transformed into two values, that is, 0 (under-expressed) and 1 (over-expressed). Only two gene expression levels are used because of the small size of the training dataset. There are only 38 training samples and the inevitable data sparseness problem can be smoothed by the small number of discretization levels. There are several discretization methods [Dougherty et al., 1995] and the following method was used. Each gene expression level was divided into under-expressed (0) and over-expressed (1) on the basis of its mean expression level across the training samples.

There are 7,129 genes and this number is too large to be applied directly to a general Bayesian network learning algorithm. Moreover, the small number of training samples available makes the quantitative part (local probability distributions) of the learned Bayesian networks unreliable. In order to learn a reliable Bayesian network from such a small dataset, an appropriate number of genes that are significantly correlated to the cancer classification should be selected. To accomplish this from only 38 training samples, we selected four genes that are significantly correlated to the leukemia class. First, the mutual information value between the leukemia class label and each gene was used to select genes. The mutual information $I(X;Y)$ measures the amount of information that a random variable X contains about another random variable Y [Cover and Thomas, 1991] and is calculated as

$$I(X;Y) = \sum_{x,y} p(x,y) \log \frac{p(x,y)}{p(x)p(y)}. \qquad [2]$$

Here, $p(\cdot)$ denotes the empirical probability estimated from the training dataset and Σ represents the summation over the entire configurations of X and Y. But, the selected four genes with high mutual information values were all highly expressed only in AML and the Bayesian network which consists of these genes showed poor classification performance.

So, two genes were selected from the genes that are highly expressed in ALL and another two genes were selected from the genes that are highly expressed in AML. For this, P-metric in [Slonim et al., 2000] was used as well. P-metric is defined as follows:

$$P(g,c) = \frac{\mu_1 - \mu_2}{\sigma_1 + \sigma_2}, \qquad [3]$$

where c is the class vector and g is the expression vector. μ_1 and μ_2 are within-class mean expression levels in class 1 and class 2, respectively. σ_1 and σ_2 are standard deviations of expression levels within class 1 and 2, respectively. Table 1 shows four selected genes and their mutual information values and P-metric values.

With the selected four genes in Table 1, a Bayesian network was constructed from the training data. The Bayesian network learning procedure consists of two parts. The first part is for structural learning. This part consists of searching for the network structure that best fits the training data. To formalize the goodness of fit of a network structure with respect to the training data, some scoring metric can be used.

Table 1. Four selected genes based on the mutual information value and the *P*-metric value. The GenBank accession number is from http://www.genome.wi.mit.edu/MPR/genes-on-hu6800.html.

	Gene description	*GenBank accession number*	*MI value*	*P-metric value*
Highly expressed in ALL	C-myb gene extracted from Human (c-myb) gene, complete primary cds, and five complete alternatively spliced cds	U22376_cds2	0.37707	1.33931
	MB-1 gene	U05259_rna1	0.31631	1.10318
Highly expressed in AML	Zyxin	X95735	0.70425	-1.40577
	Leukotriene C4 synthase (LTC4S) gene	U50136_rna1	0.50185	-1.421708

There are several such scoring metrics and they are known to be asymptotically the same. Among them, the BD (Bayesian Dirichlet) metric [Heckerman et al., 1995] was used in the experiments. This metric is defined as follows:

$$p(D,B) = p(B)\cdot\prod_{i=1}^{n}\prod_{j=1}^{q_i}\frac{\Gamma(\alpha_{ij})}{\Gamma(\alpha_{ij}+N_{ij})}\prod_{k=1}^{r_i}\frac{\Gamma(\alpha_{ijk}+N_{ijk})}{\Gamma(\alpha_{ijk})}, \quad [4]$$

where D is the training data and B is a network structure, n is the number of nodes, q_i is the number of possible configurations of parents of node i, and r_i is the number of possible values of node i. α_{ijk} is the Dirichlet prior and encodes the prior knowledge about the domain. Uninformative prior value in [Heckerman et al., 1995], $\alpha_{ijk} = 1$, was used in the experiments. N_{ijk} is the number of cases in the training data D that node i has its kth value and its parents have their j^{th} configuration. α_{ij} and N_{ij} are calculated as

$$\alpha_{ij} \equiv \sum_{k=1}^{r_i}\alpha_{ijk} \qquad N_{ij} \equiv \sum_{k=1}^{r_i}N_{ijk}. \quad [5]$$

$p(B)$ is the prior score for the network structure B and the same value was assigned for all the structures. $\Gamma(\cdot)$ is the gamma function which satisfies $\Gamma(x + 1) = x\Gamma(x)$ and $\Gamma(1) = 1$. In general, the search space for the structural learning is extremely large. For a Bayesian network with n variables, the size of the search space is about $n! \times 2^{n(n-1)/2}$. However, the Bayesian network in our experiments has only five nodes (four nodes for gene expression levels and one node for the leukemia class) and it is possible to find out the best structure in the entire search space. We also restricted the number of parents of each node to be less than three in the structural search procedure to get reliable local probability distributions with such a small dataset.

The second part of Bayesian network learning is parametric learning for the local probability distributions. With Dirichlet distribution for the parameters of the unrestricted multinomial distribution, the parametric learning is resolved by a simple calculation [Heckerman, 1996]. Figure 2 is the best Bayesian network structure containing the four genes shown in Table 1. It takes a few minutes to construct this Bayesian network on a Pentium II machine.

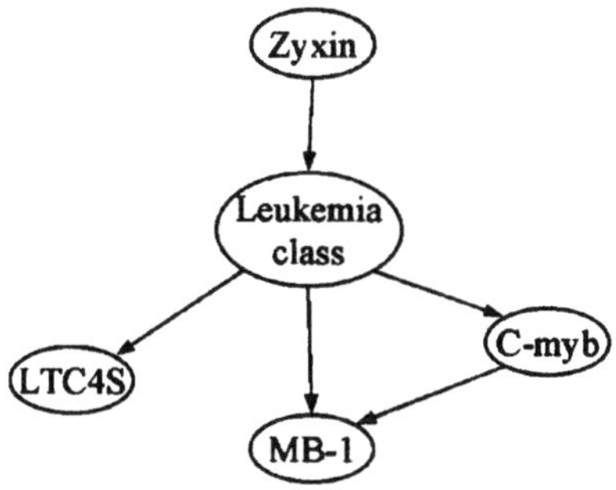

Figure 2. The Bayesian network structure with four genes shown in Table 1. An edge in the Bayesian network represents the probabilistic dependency between the child node and its parent node. This indicates the possibility of causal relationships between the child and parent. The local probability distribution for each node is not shown here.

To classify the cancer, the conditional probability *P*(Leukemia class | C-myb, MB-1, Zyxin, LTC4S) is inferred from the Bayesian network shown in Figure 2. The probability of 0.5 is regarded as misclassification.

Experimental Results for Bayesian Networks

The Bayesian network in Figure 2 correctly classifies all training samples and misclassifies only two test samples. The misclassified two samples in the test dataset are the 66th and 67th samples. Also, the network structure represents the probabilistic relationships among the four genes and the leukemia class. In Figure 2, an edge denotes the direct probabilistic dependency between the child node and its parent node. For example, "Leukemia class" depends on "Zyxin" probabilistically and vice versa. In addition, this probabilistic dependency indicates the possibility of causal dependency. It is useful to analyze gene expression patterns by Bayesian network learning to facilitate causal analysis.

MINING WITH NEURAL TREES

Neural Trees

Neural tree models [Zhang, 1994] represent multilayer feedforward neural networks as tree structures. They have heterogeneous neuron types in a single network, and the connectivity of the neurons is irregular and sparse [Zhang et al., 1997]. There are two types of neurons used in typical neural trees. One is the Σ neuron that computes the weighted sum of inputs. The other is the Π neuron that computes the product of weighted inputs. A neural tree approximates the functional relationship between inputs and outputs by its structure and weights. The advantage of neural trees over conventional neural networks is their flexibility. Neural trees can represent more complex relationships than neural networks and permit structural learning and automatic feature selection. Figure 3 is an example neural tree. In this figure, x_i is the value of the ith input node and w_{ij} is the connection weight to node i from its jth input.

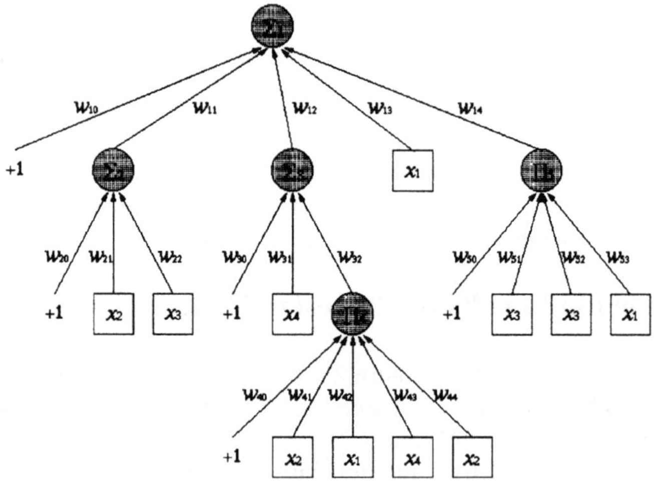

Figure 3. An example neural tree. This neural tree has four input variables (x_1, x_2, x_3, x_4). The terminal nodes denote input nodes. Note that some input nodes appear more than twice. w_{ij} is the synaptic weight to node i from its jth input and +1 is the bias to the neuron. Non-terminal nodes represent Σ neurons and Π neurons and the top-level circular node (node 1) denotes the output of the neural tree.

Evolutionary algorithms such as genetic programming can be used to find the appropriate structure and weights of neural trees. The Bayesian evolutionary algorithm (BEA) [Zhang and Cho, 2001] was used in the experiments. BEA allows the background knowledge about the given data, which is represented by the prior probability, to be incorporated in the evolutionary procedure. Thus a local search can be performed more effectively without overfitting neural trees to the training data.

Applying Neural Trees to the Cancer Classification Problem

All gene expression levels are fed to the neural tree as input. The output of the neural tree corresponds to the leukemia class label (ALL or AML). Through a structural learning process, essential genes for cancer classification are included into the neural tree and less important genes are weeded out automatically. Moreover, contrary to the Bayesian network, there is no need to discretize the inputs for neural trees. In many cases, discretization reduces information contained in the original data. In neural tree learning, all gene expression levels were linearly scaled into the interval [0.01, 0.99]. For the output value of neural tree learning, ALL was set to 0.01 and AML was set to 0.99.

The Bayesian evolutionary algorithm (BEA) [Zhang and Cho, 2001] was used to construct neural trees from training samples. It took about 10 hours to learn a neural tree from training data on a Pentium II machine.

Experimental Results for Neural Trees

Ten neural trees were constructed by BEA and their performance is summarized in Table 2. Because the evolutionary process for neural tree learning is a probabilistic search, the learned neural trees can be different from each other. Experimental results show that training samples can always be separated into two classes perfectly. Test samples are also separated well with only one misclassification for the best case although the average number of misclassified test samples is about 5. The only misclassified sample in the best case was the 66th sample.

The best model found in the second run uses just 16 genes among 7,129 genes to classify cancer types. The other genes were automatically weeded out through the evolutionary process used in neural tree learning. Table 3 shows these 16 genes. Among the 16 genes, only "Zyxin" and "GB DEF = Homeodomain protein HoxA9 mRNA" were found by other gene selection methods [Golub et al., 1999; Slonim et al., 2000].

Table 2. Prediction accuracy of neural trees. Training error is the regression error and nearly zero in all cases. The neural tree found in the second run shows the best classification performance on the test dataset.

Run	*Training error*	*Test error*
1	5.09E-04	5/34
2	4.53E-04	**1/34**
3	4.84E-04	9/34
4	8.13E-04	5/34
5	4.37E-04	8/34
6	6.35E-04	2/34
7	5.40E-04	4/34
8	4.83E-04	8/34
9	5.69E-04	2/34
10	7.43E-04	5/34
Average	5.67E-04	4.9/34

Table 3. The 16 genes found by the best neural tree. Only "Zyxin" and "GB DEF = Homeodomain protein HoxA9 mRNA" are also found by other gene selection methods. With these 16 gene expression levels, the neural tree classifies all the training samples correctly and misclassifies only one test sample. The GenBank accession number in parentheses is from http://www.genome.wi.mit.edu/MPR/genes-on-hu6800.html.

Type I keratin, hHa5 (X90763)	**Zyxin (X95735)**
GB DEF = Homeodomain protein HoxA9 mRNA (U82759)	MST1R Protein-tyrosine kinase RON (X70040)
Fibroblast Growth Factor Receptor K-Sam, Alt. Splice 1 (HG3432-HT3618)	Clone 23803 mRNA, partial cds (U79298)
PROBABLE UBIQUITIN CARBOXYL-TERMINAL HYDROLASE (D29956)	MYL1 Myosin light chain (alkali) (M31211)
LEUKOCYTE ELASTASE INHIBITOR (M93056)	Nuclear Factor Nf-Il6 (HG3494-HT3688)
GB DEF = Cdc5, partial cds (D85423)	KIAA0159 gene (D63880)
LTB4H Leukotriene B4 omega hydroxylase (cytochrome P450, subfamily IVF) (D12620)	ATP6A1 ATPase, H+ transporting, lysosomal (vacuolar proton pump), alpha polypeptide, 70kD, isoform 1 (L09235)
RING protein (Y07828)	VIL2 Villin 2 (ezrin) (X51521)

MINING WITH RADIAL BASIS FUNCTION (RBF) NETWORKS

RBF Networks

Radial basis function (RBF) networks have a similar structure to that of neural networks, but their hidden neurons contain RBFs, a statistical transformation based on a Gaussian distribution. Figure 4 represents an RBF network structure. In Figure 4, x_i is the ith input node and z_i is the ith hidden node whose output value is computed as

$$z_i = \exp\left(-\frac{\|\mathbf{c}_i - \mathbf{x}\|^2}{w_i^2}\right) \qquad [6]$$

Here, $\mathbf{c}_i$ is the m-dimensional center of z_i, w_i is the width of z_i, and $\mathbf{x}$ is the m-dimensional input vector fed to z_i. y is the output of an RBF network and computed by the following equation:

$$y = \sum h_i z_i + h_0. \qquad [7]$$

RBF networks are useful for local approximations to nonlinear input-output mapping [Haykin, 1999]. The resource-allocating network (RAN) algorithm [Platt, 1991] has been used for learning the RBF network and the appropriate number of hidden neurons is automatically determined through the learning process in this algorithm. In our experiments, the active RAN algorithm, the modified version of the RAN algorithm for the active learning scheme was used [Park and Zhang, 2000].

Applying RBF Networks to the Cancer Classification Problem

An appropriate number of genes are selected based on the P-metric [Slonim et al., 2000] and mutual information values. And then, these gene expression levels are fed into the RBF network as input. Similar to the case of neural tree learning, all input gene expression levels were linearly scaled into the interval [0, 1]. The output value of an RBF network corresponds to the leukemia class label and was set to 0 for ALL and 1 for AML.

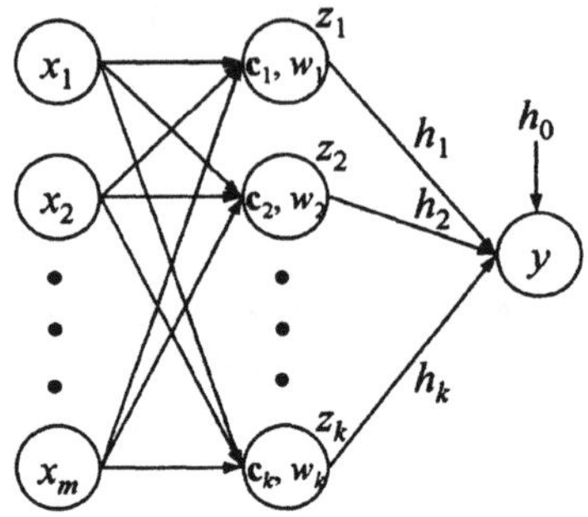

Figure 4. A radial basis function (RBF) network. ($x_1, x_2, \ldots, x_m$) is the m-dimensional input vector. $z_1, z_2, \ldots, z_k$ are k hidden neurons. The number of hidden neurons are determined automatically through the learning process. $\mathbf{c}_i$ and w_i are the center and the width of the hidden neuron z_i, respectively. h_i is the weight of z_i and h_0 is the bias to the output y.

Experimental Results for RBF Networks

Experiments were performed on 3 different sets of input genes. First, all 7,129 genes were used. Second, 50 gene expression levels by P-metric [Slonim et al., 2000] were used. On the third experiment, 10 genes were selected by mutual information and used for learning.

The active RAN (ARAN) algorithm [Park and Zhang, 2000] was used to construct RBF networks from training samples. In the case of using all 7,129 gene expression levels as an input, it took about 12 hours to learn an RBF network on a Pentium II machine. In the second and third experiments where 50 and 10 genes were used for training, it took a few minutes to construct an RBF network. Because the active RAN algorithm has the probabilistic property, its result may be different according to each run. So, we ran the active RAN algorithm on the second and the third set of genes ten times respectively.

Table 4 shows the result of each experiment. The RBF network with 50 gene expression levels classifies all training samples correctly and its test error is 1.3/34 in average.

Table 4. The prediction accuracy of RBF networks. The RBF network constructed with 50 genes selected by the P-metric value shows the best performance.

	All genes	*50 genes*	*10 genes*
No. of runs	1	10	10
Average training error	0/38	0/38	0/38
Average test error	4/34	**1.3/34**	2.5/34

COMPARISON AND EVALUATION OF THE RESULTS OF MACHINE LEARNING TECHNIQUES ON THE CANCER CLASSIFICATION PROBLEM

We compared the three classes of machine learning techniques to analyze gene expression data for cancer classification. All three methods were able to classify all training examples correctly. For test samples, neural trees and RBF networks showed almost perfect prediction accuracy. Only one test sample was misclassified. The misclassified sample is the 66th sample. The best Bayesian network classifier misclassified two test samples, the 66th and 67th samples. Table 5 summarizes the performances of three machine learning techniques.

Table 5. Comparison of the performance of three machine learning techniques. This table shows the best prediction performance of each method. The 66th sample is misclassified in all three methods.

	Training error	*Test error*
Bayesian networks	0/38	2/34
Neural trees	0/38	**1/34**
RBF networks	0/38	**1/34**

Although all three methods show nearly perfect prediction accuracy, they have some different characteristics. Neural trees and RBF networks are regression models and their inputs are numerical values. They can deal with thousands of genes. Moreover, neural trees can select some genes that are important for automatic cancer classification through its evolutionary procedure. Bayesian networks are a kind of probabilistic graphical models and their outputs are the probability of ALL (or AML) given some gene expression levels. To learn a reliable Bayesian network from the small dataset, all gene expression levels were discretized and only four genes were selected.

In order to find significantly correlated genes to the cancer classification problem, several methods were tested, including *P*-metric, mutual information, and neural tree learning. The set of selected genes based on the *P*-metric value and that based on the mutual information value were similar. On the contrary, neural trees found somewhat different genes compared with the other two methods.

In addition to the selection of essential genes, the mining of the underlying relationships among genes is important in bioinformatics. Among three machine learning techniques, Bayesian network learning has the power to capture the relationships among genes in comprehensible format. Neural tree learning seems the best in finding out a small set of interesting genes for effective classification.

Table 6 summarizes the comparative characteristics of machine learning techniques we used in the experiments.

Table 6. Comparative advantages of the three machine learning methods. It should be mentioned that this evaluation is relative and confined to the problem of cancer classification with gene expression levels we addressed.

	Bayesian networks	*Neural trees*	*RBF networks*
No. of manageable genes	Small	Large	Large
Learning time	Short	Long	Short
Classification performance	Good	Excellent	Excellent
Finding significant genes	No	Yes	No
Finding probabilistic relationships	Yes	No	No

CONCLUSION AND FUTURE WORK

In this paper, we applied three machine learning techniques, i.e., Bayesian networks, neural trees, and radial basis function (RBF) networks, to the cancer classification problem. We analyze their performance and characteristics in three different aspects.

The first criterion is the prediction accuracy. The number of training samples in the CAMDA Data Set 2 is only 38, while the number of genes is 7,129. So, the data sparseness problem is inevitable. According to the characteristics of each learning method, all or some gene expression levels were used as inputs. It is interesting to see that in spite of data sparseness, we can get satisfactory prediction accuracy. All three methods classify all training samples correctly and misclassify only one or two test samples. The only misclassified sample in all the methods was the 66th sample and this sample is thought to have some different characteristics from other samples.

The second issue is the selection of significantly correlated genes to the classification of cancer. This point of view is of course closely related to the prediction accuracy. To resolve the problem of data sparseness, the selection of some appropriate genes is necessary. We applied the *P*-metric value [Slonim et al., 2000], mutual information value [Cover and Thomas, 1991], and neural tree learning [Zhang and Cho, 2001] to select genes significantly correlated to cancer classification. Neural trees can automatically select appropriate genes in their learning procedure and genes found by neural tree learning are somewhat different from genes selected by the other two methods. It is interesting that there are some different subsets of genes with which the classification of leukemia is possible. "Zyxin" is selected by all three methods and thought to play a key role in the classification of acute leukemias.

The last point of view is capturing the relationships among gene expression levels. Bayesian networks have the ability to capture probabilistic relationships among gene expression levels in human comprehensible format. These probabilistic relationships are represented by edges in Bayesian networks and can be interpreted as the indications of possible causal relationships among gene expressions. Of course, the real causal relationships among gene expressions can only be revealed by further biological experiments and analyses. However, the learning of Bayesian networks may give some draft view of the possible causal relationships among genes and help the design of further biological experiments. Neural trees can also represent some functional relationships among gene expression levels but the relationships are hard to understand and hardly represent causal relationships.

As a conclusion, neural trees and RBF networks are very good classifiers. They show the excellent prediction accuracy despite the data sparseness inherent in cancer classification with gene expression levels. Bayesian network learning is a very good approach for the analysis of gene expression data because it is able to reveal probabilistic relationships among gene expression levels.

In our experiments, gene expression levels were discretized for Bayesian network learning. The applied method is very simple and more sophisticated discretization by human experts or combining discretization into the Bayesian network learning process [Friedman and Goldszmidt, 1996] will improve the performance of Bayesian networks. Data sparseness makes the construction of a reliable Bayesian network difficult. Because of the small sample size, only four genes were used to learn the Bayesian network. In the case of concentrating only on the qualitative part (structure) of the Bayesian network, the statistical methods such as bootstrap can be useful to induce a better network structure from data [Friedman et al., 1999; Friedman et al., 2000]. Methods for learning the more reliable quantitative part (local probability distributions) of Bayesian networks from such a small dataset should be studied further.

ACKNOWLEDGEMENTS

This work was supported in part by BK21-IT Program and the Brain Science and Engineering Program.

REFERENCES

Chickering, D.M., Learning Bayesian networks is NP-complete, In D. Fisher and H.-J. Lenz, editors, *Learning from Data: Artificial Intelligence and Statistics V*, pp. 121-130, Springer-Verlag, 1996.

Cover, T.M. and Thomas, J.A., *Elements of Information Theory*, John Wiley & Sons, 1991.

Dougherty, J., Kohavi, R., and Sahami, M., Supervised and unsupervised discretization of continuous features, In *Proceedings of the 12th International Conference on Machine Learning (ICML 1995)*, pp. 194-202, 1995.

Friedman, N. and Goldszmidt, M., Discretizing continuous attributes while learning Bayesian networks, In *Proceedings of the 13th International Conference on Machine Learning (ICML 1996)*, pp. 157-165, 1996.

Friedman, N. and Goldszmidt, M., Learning Bayesian networks with local structure, In M.I. Jordan, editor, *Learning in Graphical Models*, pp. 421-459, MIT Press, 1999.

Friedman, N., Goldszmidt, M., and Wyner, A., Data analysis with Bayesian networks: a bootstrap approach, In *Proceedings of the 15th Conference on Uncertainty in Artificial Intelligence (UAI 1999)*, pp. 196-205, 1999.

Friedman, N., Linial, M., Nachman, I., and Pe'er, D., Using Bayesian networks to analyze expression data, In *Proceedings of the 4th Annual International Conference on Computational Molecular Biology* (*RECOMB 2000*), pp. 127-135, 2000.

Golub, T.R. et al., Molecular classification of cancer: class discovery and class prediction by gene expression monitoring, *Science*, vol. 286, no. 5439, pp. 531-537, 1999.

Haykin, S., *Neural Networks: A Comprehensive Foundation*, 2nd edition, Prentice Hall, 1999.

Heckerman, D., Geiger, D., and Chickering, D.M., Learning Bayesian networks: the combination of knowledge and statistical data, *Machine Learning*, vol. 20, no. 3, pp. 197-243, 1995.

Heckerman, D., A tutorial on learning with Bayesian networks, Technical Report MSR-TR-95-06, Microsoft Research, 1996.

Jensen, F.V., *An Introduction to Bayesian Networks*, University College London Press, 1996.

Park, S.-W. and Zhang, B.-T., Learning constructive RBF networks by active data selection, In *Proceedings of the 7th International Conference on Neural Information Processing* (*ICONIP 2000*), pp. 1411-1415, 2000.

Pearl, J., *Probabilistic Reasoning in Intelligent Systems: Networks of Plausible Inference*, Morgan Kaufmann, 1988.

Platt, J., A resource-allocating network for function interpolation, *Neural Computation*, vol. 3, no. 2, pp. 213-225, 1991.

Slonim, D.K., Tamayo, P., Mesirov, J.P., Golub, T.R., and Lander, E.S., Class prediction and discovery using gene expression data, In *Proceedings of the 4th Annual International Conference on Computational Molecular Biology* (*RECOMB 2000*), pp. 263-272, 2000.

Zhang, B.-T., Effects of Occam's razor in evolving sigma-pi neural nets, In Y. Davidor, H.-P. Schwefel, and R., Männer, editors, *Lecture Notes in Computer Science*, vol. 866, pp. 462-471, Springer-Verlag, 1994.

Zhang, B.-T., Ohm, P., and Mühlenbein, H., Evolutionary induction of sparse neural trees, *Evolutionary Computation*, vol. 5, no. 2, pp. 213-236, 1997.

Zhang, B.-T. and Cho, D.-Y., System identification using evolutionary Markov chain Monte Carlo, *Journal of Systems Architecture*, vol. 47, no. 7, pp. 587-599, 2001.

Glossary

Bayesian network – A kind of graphical model that assumes a directed acyclic graph (DAG) structure.

beam search – An algorithmic search that proceeds by evaluating multiple solutions in parallel and selecting the best ones.

binary expression tree – A data structure in computer science that organizes functions or data in a tree.

biological replicates – For microarray gene expression experiments, RNA obtained from different samples of the same biological type.

cross-validation – The use of a portion of a dataset for developing a model and a remaining independent portion of a dataset for evaluating the model.

clustering analysis – Classification of objects into groups based on values of a set of variables.

data mining – Data mining is concerned with the automatic or semiautomatic exploration and analysis of large quantities of data in order to discover meaningful, previously unknown, non-trivial, and potentially useful knowledge in terms of patterns, trends, rules, relationships.

evolutionary computation – Computation inspired by the principles of evolution by natural selection.

Glossary – cont.

exhaustive search – An algorithmic search that proceeds by evaluating every possible solution to a problem.

false-discovery rate – The proportion of total predictions which are false (false positives divided by total predictions.)

feature selection - Finding *m* features (variables) of the original *N* features that are most meaningful for the classification.

fitness landscape – The relationship between statistical model variability and model quality. For smooth fitness landscapes, similar statistical models have similar quality. For rugged fitness landscapes, similar models may have differenting qualities.

genetic programming – An approach to machine learning that is based on simulated evolution, in which hypotheses are described by computer programs.

genetic algorithm - A computer algorithm that mimics Darwin's theory of evolution.

graphical model – The graph in which nodes represent random variables and the lack of arcs represents conditional independence assumptions.

hill climbing search – An algorithmic search that proceeds by evaluating a single solution, changing it, and keeping it if better than the previous solution.

Linear Discriminant Analysis (LDA)- A linear classifier based on the estimations of means and covariance matrices.

machine learning – Machine learning is concerned with the automated acquisition of and discovery of domain-specific knowledge, and with computer programs that automatically improve their problem-solving capabilities through experience. Machine learning usually involves searching a very large space of possible hypotheses to determine one that best fits the observed data and any prior knowledge held by the learner.

mutual information – The measurement of the amount of information that a random variable contains about another random variable.

Glossary – cont.

neural tree – The flexible representation of an artificial neural network as a tree structure.

non-deterministic differential expression – Gene expression intensity distributions that overlap to some degree yet have different means.

non-parametric methods – Statistical methods that make minimal assumptions about the form of the distributions involved.

NP-hard – The complex class of decision problems that are intrinsically harder than those that can be solved by a nondeterministic Turing machine in polynomial time.

optimal hyperplane - A hyperplane that separates two classes of data points and maintains a maximal margin between the two classes.

Principal Component Analysis (PCA)- Reducing the number of variables using linear combinations of the original variables while preserving the most important characteristics of the data structure. The first principal component (PC) explains the largest variation in the data.

Radial Basis Function (RBF) network – A type of artificial neural network constructed from spatially localized kernel functions.

supervised pattern recognition – Use of learning or training objects to derive a classification rule that is subsequently used to classify objects of unknown origin.

Support Vector Machine (SVM) - A classifier based on an optimal hyperplane in a feature space. Nonlinearity is introduced through a kernel function that implicitly defines a mapping from the input space to the feature space and an inner product in the feature space.

unsupervised pattern recognition - Classifying objects into classes that are not known beforehand, e.g. clustering analysis.

Index

MIX
Papier aus verantwortungsvollen Quellen
Paper from responsible sources
FSC® C105338

If you have any concerns about our products,
you can contact us on
ProductSafety@springernature.com

In case Publisher is established outside the EU,
the EU authorized representative is:
Springer Nature Customer Service Center GmbH
Europaplatz 3, 69115 Heidelberg, Germany

Printed by Libri Plureos GmbH
in Hamburg, Germany